Tantalum and Niobium-Based Capacitors

Yuri Freeman

Tantalum
and Niobium-Based
Capacitors

Science, Technology, and Applications

Second Edition

 Springer

Yuri Freeman
Greer, SC, USA

ISBN 978-3-030-89516-7   ISBN 978-3-030-89514-3   (eBook)
https://doi.org/10.1007/978-3-030-89514-3

This Springer imprint is published by the registered company Springer Nature Switzerland AG
The registered company address is: Gewerbestrasse 11, 6330 Cham, Switzerland

To my wife, Galina,
for all the good she brought into my life.

Preface

Tantalum capacitors are broadly used in modern electronics due to their record high charge and energy efficiency and capacitance stability over temperature and bias voltage. In some applications, niobium-based capacitors can be used as substitutes for tantalum capacitors. Y. Freeman's book *Tantalum and Niobium-Based Capacitors*, published by Springer in 2017, is the only book dedicated to the science, technology, and applications of these capacitors. This book received excellent feedback from the industry and academia. Since the publication of the first edition, principally important new results on the impact of technology on the reliability, failure mode, volumetric efficiency, and environmental stability of Solid Electrolytic and Polymer Tantalum capacitors, which dominate the market, were obtained. Based on these results, new possibilities for the reliable mission critical applications of the surface mount tantalum capacitors manufactured with advanced technologies were demonstrated. These new results not only significantly expand the scope of the book but also provide important corrections and clarity to the earlier published material. The expanded second edition of the book *Tantalum and Niobium-Based Capacitors: Science, Technology, and Applications* with these additions, corrections, clarifications, and new application opportunities will be of great interest to the developers, makers, and users of tantalum and niobium-based capacitors in related industries as well as electrical engineering and material science faculty and university students.

Greer, SC, USA   Yuri Freeman

Acknowledgments

I would like to honor the memory of the great scientists with whom I was privileged to work alongside in my career: Prof. Lev Palatnik from Kharkiv Polytechnic Institute; Dr. Ilia Skatkov from SCB ELITAN; Dr. Felix Zandman, founder of Vishay Intertechnology; and Dr. Brian Melody from KEMET Electronics.

I would like to acknowledge Dr. Terry Tripp, whose papers and later collaboration encouraged my work in the field.

I would like to thank my colleagues in the industry and academia who contributed to this book directly or through valuable conversations: KEMET's Steve Hussey, Dr. Phil Lessner, Jonathan Paulsen, Tony Kinard, Dr. James Chen, George Haddox, Barry Reeves, Dr. Erik Reed, Dr. Javaid Qazi, Angela Parker, Jeff Poltorak, Jim Fife, David Jacobs, Paul Young, Chris Guerrero, John Ols, Steve Brooks, Cynthia Prince, Kevin Hanley, Henry Bishop, Kirby Smith, Traci McAdam, Joshua Reid, and Craig Busby; Clemson University's Dr. Rod Harrell, Dr. Igor Luzinov, and Dr. Ruslan Burtovyy; and Tel Aviv University's Dr. Alexander Gladkikh, Dr. Yuri Rozenberg, and Dr. Alexander Palevski.

I would also like to acknowledge the engineers, technicians, and operators in the manufacturing, analytical, and electrical labs and mechanical shops who helped transform scientific ideas into advanced technologies and products.

I would like to thank KEMET managers Chuck Meeks, Ed Jones and Hal Perkins for their support and help with promoting novel tantalum capacitors into the most conservative military and aerospace markets.

Special thanks to my children Anya and Daniel for their inspiration and help with editing this book.

Contents

About the Author

Yuri Freeman received his master's degree as engineer-physicist from the renowned school of Thin Solid Films at the Kharkiv Polytechnic Institute in Ukraine. He graduated among the first in his class and hoped to continue his work in academia. However, as it was still Soviet Union times and graduates were "distributed" often contrary to their choices, Yuri was sent to work for SCB ELITAN, a developer and producer of tantalum capacitors. At first Yuri thought sadly: "What could be more primitive than a capacitor with two plates separated by a dielectric?" He realized only later that his random assignment was an incredible opportunity for a scientist and an engineer! The change started with his involvement in the development of niobium capacitors to substitute tantalum capacitors, enforced by a shortage in tantalum in former Soviet Union. Comparison of the degradation mechanisms in tantalum and niobium capacitors became the topic of his Ph.D. in solid-state physics.

Shortly after the collapse of the Soviet Union, SCB ELITAN was closed and Yuri started his work for **Sprague-Electric** later Vishay-Sprague, the original manufacturer of solid tantalum capacitors. After several years with Vishay Sprague, where Yuri was promoted to principal scientist, his facility in Sanford, Maine, USA, was closed, and manufacturing moved overseas. While Yuri was looking for a new job, his Siberia-born wife fell in love with sunny South Carolina, and that's how Yuri ended up at KEMET Electronics, a global producer of tantalum capacitors. This turned out to become the best part of his career with the most challenging projects and the industry's best team to work with. Now, Yuri is a Fellow/VP, and director of strategic development in tantalum at KEMET Electronics. As an adjunct professor, he is also teaching science and technology of the electronic components at the Clemson University.

In 2018, the Tantalum and Niobium International Study Center (TIC) awarded Yuri Freeman the Ekeberg Prize for the "outstanding contribution to the advancement of the knowledge and the understanding of the metallic elements tantalum and niobium."

Introduction

According to Greek mythology, Tantalus was a Greek king who held the friendship and favor of the gods until he challenged their will and was consigned to perpetual thirst and hunger. The name Tantalum was given to the metal element with atomic number 73 and symbol Ta by a Swedish chemist A. Ekeberg who discovered it in 1802 in minerals taken from Kimito, Finland, and Ytterby, Sweden. The custom of the time was to dissolve new elements in mineral acids to investigate their chemical properties. Yet this was tantalizingly difficult for Ekeberg to accomplish since tantalum was exceptionally unaffected by mineral acids. Later research revealed that tantalum owes its stability in aggressive environments to a thin film of native oxide that covers tantalum surface when the metal is exposed to oxygen. In addition to chemical stability, tantalum pentoxide (Ta_2O_5) thin films obtained by chemical vapor deposition, thermal oxidation of sputtered layers of tantalum, and anodic oxidation of tantalum in aqueous and nonaqueous solutions are among the best-known dielectrics, combining high dielectric constant and electrical strength [1]. According to the Tantalum-Niobium International Study Center (TIC), about a half of the tantalum produced globally is consumed in a wide variety of electronic components and circuits, utilizing exceptional dielectric properties of the Ta_2O_5 films. In these applications, Ta_2O_5 films are employed as dielectric layers for storage capacitors in DRAM, gate oxides in field effect transistors, dielectrics in tantalum capacitors, etc.

Tantalum capacitors are broadly used in modern electronics due to their stability in a wide range of temperatures and frequencies, long-term reliability, and record high volumetric efficiency CV/cm^3 (where C represents capacitance and V represents applied voltage). These capacitors can be found in the most demanding applications such as medical, aerospace, and military, as well as in numerous commercial applications such as automotive, telecommunications, and computers. Tantalum capacitors are the essential component of cardio implants, which automatically sense abnormal heart rhythms and deliver an electric countershock within seconds, saving thousands of people from sudden cardiac death [2]. Figure 1 shows a Medtronic cardio implant with tantalum capacitors as the key part of the therapy provided.

Fig. 1 Medtronic cardio
implant with tantalum
capacitors

Fig. 2 Leaded (**a**) and chip (**b**) tantalum capacitors

Tantalum capacitors consist of a tantalum anode made of tantalum powder sintered in vacuum; an anodic oxide film of tantalum employed as a dielectric; and a cathode which can be either a liquid electrolyte (Wet Tantalum capacitors), manganese dioxide (Solid Electrolytic Tantalum capacitors), or an inherently conductive polymer (Polymer Tantalum capacitors). Depending on the type of assembly and encapsulation, these capacitors can be either leaded, typically in hermetic metal cans, or surface-mount chips typically in non-hermetic plastic casing. Some surface-mount chip configurations with ceramic or plastic-over-metal casing can also be hermetic. Figure 2 illustrates examples of the leaded and chip tantalum capacitors.

Figure 2a presents leaded Wet, Solid Electrolytic, and Polymer Tantalum capacitors (from the left to the right). Figure 2b presents a variety of chip Solid Electrolytic and Polymer Tantalum capacitors.

Fig. 3 Marking on a chip
tantalum capacitor

Tantalum capacitors are polar with positive polarity on tantalum anode. The bar and/or sign "+" indicates positive termination on the marking of the capacitor. Figure 3 shows an example of the marking on a chip tantalum capacitor.

This marking also shows the type of the capacitor (KM stands for T540 series KEMET Polymer Tantalum capacitor), capacitance (337 stands for 330 µF), working voltage (16 V), tolerance (K stands for +/− 10%), and date code (710 stands for Year 2017, Week 10).

The first Wet Tantalum capacitors were manufactured in the 1930s by Fansteel Metallurgical Corporation, the largest producer of tantalum in the USA at that time [3]. Early Wet Tantalum capacitors employed anodes made of tantalum foil, which were later replaced with porous tantalum anodes sintered in vacuum with tantalum powder. These anodes had a considerably higher surface area per unit of volume in comparison to the surface area of the foil anodes, providing greater volumetric efficiency. Porous anodes sintered in vacuum with capacitor-grade tantalum powder are now used in all types of tantalum capacitors.

The disadvantage of Wet Tantalum capacitors is their high equivalent series resistance (ESR), especially at low temperatures, which causes capacitance loss and limits ripple current, operating frequencies, and the range of operating temperatures. The major contributor to ESR in Wet Tantalum capacitors is the resistance of the liquid electrolyte within the capacitor, typically an aqueous solution of sulfuric acid. Current flow through the liquid electrolyte is provided by ions with much lower mobility in comparison to mobility of current carriers in solid conductors. At lower temperatures, viscosity of the electrolyte increases and mobility of ions decreases. When ambient temperature approaches the freezing point of the electrolyte, ion mobility becomes negligible, causing extremely high ESR and total loss of capacitance.

The invention of the solid transistor by J. Bardeen, W. Brattain, and W. Bradford from Bell Telephone Laboratories in the USA, for which they received the Nobel Prize in Physics in 1956, marked the beginning of modern electronics [4]. The replacement of massive vacuum tube transistors with small semiconductor-based transistors allowed for drastic decrease in size and increase in operating frequencies, thereby improving the functionality of electronic devices.

To meet these challenges in the capacitor world, H. Haring, N. Summit, and R. Taylor from Bell Telephone Laboratories invented the Solid Electrolytic Tantalum capacitor with manganese dioxide (MnO_2) cathode, initially named the "Dry Electrolytic device" [5]. To produce MnO_2 cathode according to this invention, tantalum anodes were sintered in vacuum and anodized and then immersed in an aqueous solution of manganese nitrate, followed by pyrolysis of manganese nitrate into solid manganese dioxide and gaseous nitrogen oxide. Manganese dioxide covered the surface of the dielectric inside and outside of the porous anode forming the cathode of tantalum capacitor, while gaseous nitrogen oxide evaporated from the capacitor body. Using this technology, Sprague Electric in the USA began mass manufacturing of Solid Electrolytic Tantalum capacitors in the late 1950s. D. Smyth, G. Shirn, and T. Tripp, three leading scientists at Sprague Electric, made significant contribution to the science and technology of Solid Electrolytic Tantalum capacitors [6–12].

The major advantage of Solid Electrolytic Tantalum capacitors in comparison to Wet Tantalum capacitors is their low ESR. This is because MnO_2 is an n-type semiconductor with electrons as majority current carriers having much higher mobility compared to the mobility of ions in liquid electrolyte in Wet Tantalum capacitors. Next, MnO_2 has a narrow band gap, providing a significantly smaller temperature dependence of ESR versus the temperature dependence of ESR in Wet Tantalum capacitors. Similar to the liquid electrolyte cathode in Wet Tantalum capacitors, MnO_2 is an oxidizer releasing oxygen into the oxide dielectric to compensate oxygen vacancies there. Manganese dioxide cathodes also provide Solid Electrolytic Tantalum capacitors with a strong self-healing capability: MnO_2 can transform locally into lower manganese oxides such as Mn_2O_3, Mn_3O_4, and MnO with considerably higher resistivity than resistivity of the MnO_2 cathode. This transformation occurs in defect areas of the dielectric, which have higher current density and, thereby, higher temperatures. The low manganese oxides with high resistivity block current flow through the defects in the dielectric, preventing thermal runaway and capacitor failure.

The conductive polymer cathode was developed in the 1980s and 1990s by NEC Corporation in Japan. Conductive polymer cathode was initially developed for aluminum capacitors and later applied to tantalum capacitors to satisfy demand for miniaturization and high operating frequencies by the fast-growing computer and telecommunication industries [13]. NEC was also the first mass manufacturer of Polymer Tantalum capacitors. Poly(3,4-ethylenedioxythiophene) (PEDOT) is commonly used as the cathode material in Polymer Tantalum capacitors [14–16].

The major advantage of Polymer Tantalum capacitors in comparison to Wet and Solid Electrolytic Tantalum capacitors is their low ESR with reduced dependence on temperature. This is because the conductivity of the inherently conductive polymer is in the range of 100–1000 S/m, compared to the conductivity of the MnO_2, which is in the range of 0.1–1 S/m.

The history of tantalum capacitors reveals that the chief driving force in their evolution from Wet, to Solid Electrolytic, to Polymer was ESR reduction. At the same time, this evolution resulted in much lower working voltages in Solid

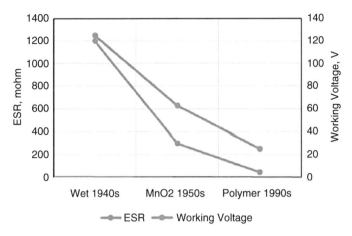

Fig. 4 ESR and maximum working voltage in different types of tantalum capacitors

Electrolytic and Polymer Tantalum capacitors in comparison to Wet Tantalum capacitors. Typically, the working voltages achieve 125 V in Wet Tantalum capacitors and 63 V in Solid Electrolytic Tantalum capacitors. When they were introduced to the market, the working voltages of Polymer Tantalum capacitors have been limited to 25 V. Figure 4 shows typical ESR and maximum working voltage in tantalum capacitors with different cathodes.

The limited working voltage has made the applications of original Polymer Tantalum capacitors impossible in high-voltage, high-reliability circuits, where the low ESR is most critical. Traditionally, the working voltage of tantalum capacitors is increased through the use of higher formation voltage during electrochemical oxidation of the tantalum anodes. Since the thickness of the anodic oxide film of tantalum is directly proportional to the formation voltage, a higher formation voltage results in a thicker Ta_2O_5 dielectric and, thereby, a greater breakdown voltage (BDV) and working voltage for tantalum capacitors [17]. However, as the formation voltage increased in original Polymer Tantalum capacitors with PEDOT cathodes, the BDV increased to approximately 50 V and then leveled off, even though the dielectric thickness continued to grow [16]. This "saturation" of the BDV limited the ultimate working voltage of original Polymer Tantalum capacitors to 25 V since many years of testing and field application of different types of tantalum capacitors established that a working voltage should be no larger than 50% of the BDV to achieve long-term stability and reliability.

Another significant disadvantage of original Polymer Tantalum capacitors was their high direct current leakage (DCL), which was typically about ten times higher than DCL in Solid Electrolytic Tantalum capacitors and even more than that in comparison to DCL in Wet Tantalum capacitors. Moreover, in many cases, DCL in original Polymer Tantalum capacitors continued to grow during life test and field applications, causing parametric and catastrophic failures of these capacitors. As a result of the combination of low BDV and high DCL, original Polymer Tantalum capacitors were used only in low-voltage commercial applications, which do not require long-term stability and reliability. A deep understanding of the degradation and conduction mechanisms in Polymer Tantalum capacitors and breakthroughs in their technology and screening techniques were needed to overcome these limitations and develop high-voltage and low-leakage Polymer Tantalum capacitors for high-reliability applications.

Niobium (Nb) was an obvious choice to substitute tantalum as the anode material in the electrolytic capacitors because the physical and chemical properties of niobium and tantalum have so much in common that for some period of time they were considered the same metal. The names of these two metals are also connected. In Greek mythology, Niobe is Tantalus's daughter. The first niobium capacitors were developed in Leningrad (now St. Petersburg) in former USSR where sources of tantalum were limited, while niobium was plentifully available. Mass manufacturing of niobium capacitors was launched in the 1960s. Later, a joint team of engineers and scientists from SCB ELITAN and Kharkiv Polytechnic Institute in Ukraine made significant contributions to the science and technology of niobium capacitors [18, 19]. OJSC ELECOND in Russia continues mass manufacturing of Solid Electrolytic Niobium capacitors for commercial and special applications.

In the West, interest in niobium capacitors was fueled by the shortage of tantalum in the early 2000s due to the vast consumption of Solid Electrolytic and Polymer Tantalum capacitors in the fast-growing telecommunication industry. Besides niobium, niobium monoxide (NbO) with a metallic-type conductivity was introduced by J. Fife as anode material for electrolytic capacitors [20]. Mass manufacturing of the Niobium Oxide capacitors with NbO anodes was launched by the AVX Corporation in 2003 [21] and also presented by Vishay Sprague [22].

Niobium-based capacitors consist of porous anodes sintered in vacuum with niobium or niobium oxide powders; anodic oxide film Nb_2O_5 as a dielectric; and MnO_2 cathodes. The advantage of niobium-based capacitors in comparison to tantalum capacitors is their lower cost per CV caused by larger availability and lower cost of niobium in comparison to tantalum. Despite the cost advantage, applications of niobium-based capacitors in modern electronics are very limited due to their lower volumetric efficiency and working voltages in comparison to tantalum capacitors. Furthermore, DCL in niobium-based capacitors is typically higher than DCL in tantalum capacitors and often increases with time during the capacitor testing and field application, causing parametric and catastrophic failures. In some early hermetic niobium capacitors developed in the USSR, DCL was increasing with time during storage even without voltage applied; however, when the seal was opened, DCL quickly fell back to its initial level. The initial hypothesis was that a sort of "evil spirit" was created inside the hermetic can during storage and then escaped from the can when the seal was opened. The "evil spirit" was the official name of this type of failures in niobium capacitors even after it was discovered that the real reason for the reversible DCL increase was redistribution of moisture inside the hermetic can and that adding a controlled amount of humidity to the can will help stabilize DCL in hermetic Solid Electrolytic Niobium capacitors.

Surprisingly, the Greek myth about king Tantalus and his daughter Niobe played a role in improvements in technology of niobium capacitors. As princess Niobe was a delicate and sensitive offspring of her sturdy father, the niobium capacitor technology, which was initially practically identical to the tantalum capacitor technology, was adjusted to be more gentle, with less thermal and mechanical shocks and additional healing steps to support the Nb_2O_5 dielectric. However, major improvements in performance and reliability of tantalum and niobium-based capacitors were achieved as a result of the deep understanding of the degradation mechanism in these capacitors.

Chapter 1
Major Degradation Mechanisms

The basic bilayer of all types of tantalum capacitors, tantalum anode, and anodic oxide film of tantalum as a dielectric is not a thermodynamically stable system. This is demonstrated by the tantalum-oxygen equilibrium diagram that does not contain two-phase equilibrium areas for pure tantalum and tantalum pentoxide (Fig. 1.1) [23, 24]. Relaxation of the Ta-Ta$_2$O$_5$ system into the thermodynamically stable state occurs through oxygen migration from Ta$_2$O$_5$ to Ta, resulting in a solid solution of oxygen in tantalum anode and oxygen vacancies in the tantalum oxide dielectric. Conductivity of the dielectric and, thereby, DCL of tantalum capacitor increase exponentially with the concentration x of oxygen vacancies in the depleted with oxygen Ta$_2$O$_{5-x}$ [6, 8, 11]. Tantalum capacitors with thinner dielectrics are more sensitive to oxygen migration since the thickness of the depleted with oxygen portion of dielectric near the Ta-Ta$_2$O$_5$ interface is comparable with the total thickness of the dielectric.

Another reason for the thermodynamic instability of tantalum capacitors is the amorphous structure of the anodic oxide film of tantalum formed on crystalline tantalum. The image of the Ta-Ta$_2$O$_5$ bilayer on Fig. 1.2 obtained by super high-resolution transmission electron microscopy (TEM) demonstrates the differences between the amorphous structure of the anodic oxide film of tantalum and crystalline structure of tantalum [25].

One can see from Fig. 1.2 that atoms in the anodic oxide film of tantalum are disordered and do not have any preferable orientation, which is inherent to amorphous structures, while atoms in the tantalum metal are ordered in a crystalline lattice. This amorphous structure of the anodic oxide film of tantalum employed as a dielectric in tantalum capacitors is of paramount importance to performance and reliability of these capacitors. Typically, the dielectrics in the tantalum capacitors are very thin (from 0.02 μm to 0.6 μm depending on the formation voltage). As a result of the low thickness, the electrical field in the dielectric film at rated voltage is on the order of several million V/cm. The disordered arrangement of atoms in the amorphous dielectric results in a high density of electron traps for the electric

Y. Freeman, *Tantalum and Niobium-Based Capacitors*,
https://doi.org/10.1007/978-3-030-89514-3_1

Fig. 1.1 Tantalum-oxygen
equilibrium diagram [23]

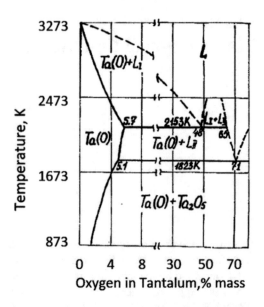

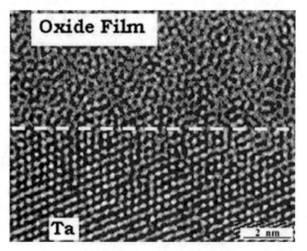

Fig. 1.2 TEM image of
the amorphous anodic
oxide film formed on
crystalline tantalum (the
white spots represent
individual atoms)

current carriers. Due to these traps, the mobility of electric current carriers stays low
in the high electrical field, preventing accumulation of critical energy that otherwise
might cause an electric avalanche and the breakdown of the dielectric.

On the other hand, the disordered structure of the amorphous dielectrics increases
the internal energy in the dielectric, making these dielectrics thermodynamically
unstable. Amorphous dielectrics trend to ordering and crystallization spontaneously
to reduce their internal energy. Growth of crystalline inclusions in amorphous
matrix of the anodic oxide film induces mechanical stress in the film due to a difference between specific volume of the amorphous and crystalline phases. Mechanical
stress associated with field-induced crystallization was detected experimentally

using in situ measurements on anodic oxide film of tantalum deposited on quartz substrate [26]. Eventually this mechanical stress results in a disruption of the dielectric and, thereby, in the failure of the capacitor. The rate of crystallization process is a principal factor governing the performance and reliability of tantalum capacitors.

The rate of the crystallization process in anodic oxide films on tantalum depends on many factors such as purity and morphology of the anode surface, type and temperature of formation electrolyte, conditions of post-formation thermal treatment, etc. In all cases, the crystallization rate increases with increasing thickness of the anodic oxide film. The effect of the dielectric thickness on the rate of crystallization can be explained by contributions to the internal energy of the dielectric by the bulk of the anodic oxide film and the interface between the anodic oxide film and anode. An amorphous film doesn't need misfit dislocations on the interface with the crystalline anode in order to adjust its structure to the crystalline lattice of the anode. Since these misfit dislocations increase internal energy, the amorphous film adjacent to the crystalline substrate allows lower internal energy than a similar crystalline film would be. This helps stabilize the amorphous state of the dielectric. On the other hand, the disordering of atoms in the bulk of the amorphous dielectric increases internal energy of the dielectric. The balance in internal energy of the dielectric and therefore the balance between the tendency of the dielectric to remain amorphous or to crystallize depend on the ratio between the surface and bulk of the dielectric. The greater is the thickness of the amorphous dielectric, the larger is the contribution of the bulk to the internal energy that makes thicker amorphous dielectrics more prone to crystallization.

Despite the inherent thermodynamic instability, tantalum capacitors can be manufactured with exceptional stability and reliability. Such high stability and reliability can be achieved because kinetics of the degradation processes, oxygen migration and crystallization, can be reduced by technological means to such a negligible level that no degradation takes place during any practical duration of the testing and field application, even at the highest possible operation temperatures and voltages. Oxygen migration and crystallization processes also take place in niobium-based capacitors, causing parametric and catastrophic failures of these capacitors. The pace of the degradation processes in niobium-based capacitors at given dielectric thickness, temperature, and applied voltage is significantly higher than that in tantalum capacitors with the same dielectric thickness, temperature, and voltage. Nevertheless, improvements in the stability and reliability of niobium-based capacitors have been also achieved by the technological means slowing down degradation processes in these capacitors.

1.1 Oxygen Migration

Comprehensive investigation of oxygen migration in tantalum capacitors was conducted during the early stages of manufacturing of Solid Electrolytic Tantalum capacitors [6–12]. The manufacturing includes multiple pyrolysis of liquid manganese nitrite impregnating porous tantalum anodes to produce a solid MnO_2 cathode. Thermal treatment associated with this process, typically performed in the range of temperatures 250–350 °C, resulted in an increase in DCL, dissipation factor (DF), and capacitance, as well as in a large capacitance dependence on bias, frequency, and temperature. Only at very low temperatures, there were practically no capacitance dependence on bias and frequency as a result of the thermal treatment.

To explain the effects of the thermal treatment on capacitance behavior of tantalum capacitors, a conductivity profile model was developed based on a gradient of oxygen vacancies in the oxide dielectric [6–8]. The conductivity profile develops during the thermal treatment as a result of oxygen migration from the tantalum oxide to the tantalum, leaving oxygen vacancies in the oxide dielectric. This transformation can be presented as:

$$Ta + Ta_2O_5 \rightarrow Ta(O)_x + Ta_2O_{5-x} + xV_o,$$

where $Ta(O)_x$ is a solid solution of oxygen in tantalum and x is a concentration of oxygen in tantalum and concentration of oxygen vacancies V_o in tantalum oxide. When the oxygen deficiency reaches the oxide-air interface, oxygen begins to diffuse through the oxide ultimately producing a gradient of oxygen vacancies each with two trapped electrons. This oxygen vacancy gradient is associated with a gradient of conductivity since electrons trapped in the oxygen vacancies can be emitted under applied voltage, increasing conductivity of the oxide dielectric. The point of highest oxygen vacancy concentration and associated conductivity is located at the metal-oxide interface, after which the level of conductivity decreases exponentially across the dielectric.

The key to the model is a concept of a critical conductivity defined as the level of conductivity, which distinguishes conducting from nonconducting oxide. During the thermal treatment, a portion of the oxide dielectric next to the tantalum interface acquires a conductivity greater than critical conductivity and can thus be considered a part of the electrode. This causes the effective dielectric thickness to decrease and the capacitance to increase. When bias voltage is applied to the dielectric, the electrons trapped in oxygen vacancies are emitted and withdrawn to the tantalum anode. This reduces conductivity in such a way to move the intersection of the conductivity profile with the critical conductivity toward the tantalum interface, thereby increasing the effective dielectric thickness and reducing the capacitance. A similar effect takes place when temperature is decreasing. In this case, the conductivity profile shifts to lower magnitudes while simultaneously shifting its intersection with the critical conductivity toward the tantalum surface. Consequently, the change in capacitance arises not only from a small change in the dielectric constant with temperature but also from the increase in the effective thickness of the dielectric.

Reformation of the dielectric at the same voltage and temperature as the initial formation eliminates oxygen vacancies in the oxide dielectric and reduces capacitance dependence on bias, temperature, and frequency to these before the thermal treatment. The magnitude of the capacitance after the reformation is equal to or slightly higher than initial capacitance before the thermal treatment. The "extra" capacitance depends on the temperature and duration of the thermal treatment and was attributed to a change in the dielectric constant of the oxide dielectric due to the structural transformations such as low-scale ordering, which takes place during the thermal treatment [9, 27, 28].

Kinetics of oxygen migration in tantalum capacitors strongly depend on the presence of dopants in the oxide dielectric. It is well known that the material from the formation electrolyte is incorporated into the outer portion of the oxide film during the anodizing process [29, 30]. Both tantalum and oxygen move across the oxide film during anodizing to form a duplex film that has a material of the electrolyte incorporated in the outer layer at a concentration that increases proportional to the concentration of the electrolyte. During thermal treatment typically performed on tantalum anodes after the initial formation, a dopant can redistribute within the tantalum oxide making it more homogenous [7]. When the formation solution is phosphoric acid, the incorporated phosphorus has a profound effect on the capacitor properties, including capacitance behavior as a result of the thermal treatment [31–33]. Particularly, a much smaller capacitance dependence on temperature was detected in the case of thermal treatment of a tantalum foil anodized in concentrated phosphoric acid in comparison with the formation in dilute phosphoric acid or other electrolyte solutions like H_2SO_4. This evidences about a slower pace of oxygen migration in doped with phosphorus tantalum oxide.

A slower pace of oxygen migration was also detected in anodized tantalum enriched with nitrogen [34–37]. Anodic oxide films produced on this tantalum have a nitrogen-doped layer at the metal side of the film. Thermal treatment in air at 350 °C for 1 h resulted in significantly smaller capacitance increase in anodized tantalum with greater than 10 atomic percent of nitrogen in comparison with anodized tantalum without nitrogen doping. It was proposed that the presence of interstitial nitrogen in crystalline lattice of tantalum greatly retards the extraction of oxygen from the anodic oxide film, and this reduces the concentration of oxygen vacancies and, thereby, conductivity of the oxide dielectric. When conductive tantalum nitride, TaN, was used as anode material for tantalum capacitors, no degradation of the anodic oxide film occurred as a result of the heat treatment at 350 °C for 1 h [38, 39]. This was indicated by no change in the capacitance and no capacitance dependence on the bias, temperature, and frequency as a result of the heat treatment.

Increase in oxygen content in tantalum anodes also reduces the pace of oxygen migration from the tantalum oxide dielectric to the tantalum anode. Oxygen content in tantalum anodes increases during the thermal treatment after the initial formation of the oxide dielectric. Reformation following the thermal treatment restores oxygen content in the dielectric; however, it doesn't affect oxygen distribution in tantalum anodes since the formation and reformation temperatures are much lower than the thermal treatment temperature. Thermal treatment of the formed tantalum anodes is usually performed in air; however, the treatment can also be performed in a vacuum or an inert atmosphere to prevent ignition when the anodes are made with fine high CV tantalum powder and have thin oxide dielectric. When there is no oxygen in an ambient atmosphere to replace oxygen which migrated from the oxide dielectric to the tantalum anode, a high concentration of oxygen vacancies develops across the oxide dielectric, and a large portion of the dielectric becomes conductive [3, 27, 32, 40].

Oxygen content in the bulk of tantalum anodes increases spontaneously during the sintering of tantalum powders due to dissolving of oxygen from the native oxide on the surface of tantalum particles prior to their sintering. This occurs at sintering temperatures equal to or below 1870 °C, which are typically used for middle and high CV tantalum powders [23]. Above 1870 °C, oxygen evaporates from tantalum to vacuum chamber during the powder sintering; however, only very coarse tantalum powders can be sintered at such high temperatures without a significant loss in surface area. Despite the fact that increased oxygen content in the bulk of tantalum anodes slows down further oxygen migration from the oxide dielectric, it can also activate the crystallization processes in the amorphous matrix of the dielectric. The effect of the bulk oxygen in tantalum anodes on the crystallization processes in anodic oxide films of tantalum formed on the surface of these anodes will be discussed in the following chapter of the book.

Experiments on anodized niobium foil revealed that the Nb-Nb_2O_5 bilayer is significantly more sensitive to thermal treatment than the Ta-Ta_2O_5 bilayer [9]. The effects of thermal treatment on capacitance and capacitance dependence on bias, temperature, and frequency observed in Ta-Ta_2O_5 system after heating in the 300–400 °C range occur in Nb-Nb_2O_5 system after heating only in the 200–300 °C range. Stronger effects of thermal treatment on anodized niobium were attributed to a greater thermodynamic instability of the Nb-Nb_2O_5 bilayer relative to the Ta-Ta_2O_5 bilayer and also to the existence of the two lower niobium oxide phases: NbO_2 with semiconductor-type conductivity and NbO with metallic-type conductivity (Fig. 1.3).

A comparison of the degradation mechanisms in solid tantalum and solid niobium capacitors with MnO_2 cathodes was presented in [18, 19]. During thermal treatment of these capacitors, two redox reactions take place: one at the Me-Me_2O_5 interface (Me: Ta or Nb) where Me anode extracts oxygen from Me_2O_5 dielectric and the other at the MnO_2-Me_2O_5 interface where MnO_2 cathode compensates oxygen in the depleted Me_2O_5 dielectric when oxygen vacancies reach this interface. These two redox reactions correspond to each other by the conduction of oxygen ions through the Me_2O_5 dielectric. Thus, under heating tantalum and niobium capacitors with MnO_2 cathodes display the properties of the galvanic cell generating electric current in the external circuit. The short circuit current (I_{sc}) generated by these capacitors at high temperatures is equal to the current of oxygen ions through the Me_2O_5 dielectric. Figure 1.4 presents I_{sc} dependence on inverse temperature in solid tantalum and niobium capacitors with MnO_2 cathodes [19].

As one can see in Fig. 1.4, the I_{sc} and, thereby, the current of oxygen ions are high in the Nb_2O_5 dielectric of the niobium capacitors at relatively low temperatures when the current of oxygen ions in the Ta_2O_5 dielectric of the tantalum capacitors is still negligibly low. Higher current of oxygen ions in the Nb_2O_5 dielectric in niobium capacitors in comparison with the current of oxygen ions in the Ta_2O_5

Fig. 1.3 Niobium-oxygen
equilibrium diagram [23]

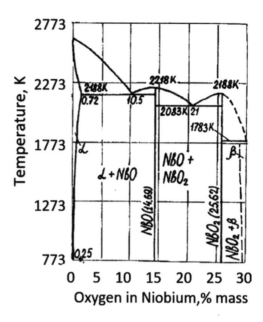

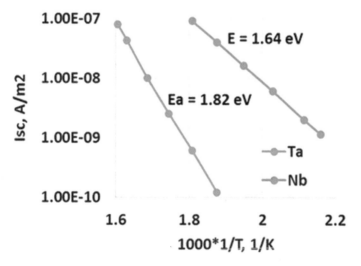

Fig. 1.4 Short circuit current I_{sc} dependence on inverse temperature in solid tantalum and niobium capacitors with MnO_2 cathodes

dielectric in tantalum capacitors corresponds to lower activation energy (E_a) of oxygen migration in these capacitors. These activation energies, calculated from $I_{sc}(T)$ data using Arrhenius equation, were $E_a = 1.64$ eV for Nb/MnO_2 capacitors and $E_a = 1.82$ eV for Ta/MnO_2 capacitors (Fig. 1.4).

A significant oxygen migration in niobium capacitors occurs in the temperature range typically used in manufacturing of these capacitors for pyrolysis of manganese nitrite to form MnO_2 cathode. At these conditions, the MnO_2 cathode becomes exhausted with active oxygen and loses the ability to compensate oxygen vacancies in the Nb_2O_5 dielectric, resulting in an increased DC leakage and DC leakage instability during testing and field application. Furthermore, at equal temperatures and dielectric thicknesses, the pace of the crystallization process in the amorphous matrix of the dielectric is also higher in the Nb_2O_5 dielectric in niobium capacitors compared to the Ta_2O_5 dielectric in tantalum capacitors.

1.2   Field Crystallization

The crystallization of the amorphous dielectric in tantalum capacitors can occur during the formation of the anodic oxide film on the surface of tantalum anode (field crystallization) and as a result of post-formation thermal treatment (thermal crystallization). D. Vermilyea investigated crystallization process during formation of anodic oxide film on tantalum foils and sputtered tantalum films [26, 27]. He showed that crystallization occurs at nucleation sites on the metal-oxide interface. These sites were attributed to the inclusion of a high impurity content in the metal. Due to the uniform impurity distribution during tantalum sputtering, the crystallization rate was much lower with sputtered tantalum films than that with tantalum foil. Vermilyea detected an incubation period for the appearance of crystalline areas in the anodic oxide film. This incubation period was associated with the time required for a crystal growing underneath the amorphous matrix to reach a critical size when a disruption of the matrix occurs due to the critical mechanical stress. That provided a direct access of the electrolyte to the tantalum surface, speeding up the crystal growth.

N. Jackson investigated the effect of formation conditions on the crystallization rate in anodic oxide films on tantalum [41]. He showed that the critical parameters controlling crystal growth are the type and temperature of the formation electrolyte. The most dramatic effect in suppressing the crystallization process was achieved with phosphoric acid. The effect of formation electrolyte, particularly the incorporation of phosphorus into the anodic oxide film during its formation in phosphoric acid, was also shown in [17, 42, 43]. B. Melody, T. Kinard, and P. Lessner demonstrated non-thickness limited anodizing under special formation conditions when anodic oxide films may be grown to practically any thickness at a constant formation voltage [44]. Obviously, these special formation conditions might provide suppression of the crystallization to grow thick anodic oxide films with amorphous structure.

Carbon and transition metals like iron, nickel, and chromium are among the impurities having the strongest effect on the density of nucleation sites for the crystallization [41, 45]. These impurities can come to the tantalum and niobium anodes from the metal powders or from materials and tooling used for powder pressing and sintering (organic binder, metallic dies, etc.). High carbon content may result in precipitation of tantalum and niobium carbides in anodes during powder sintering. The precipitates of carbide phase work as nuclei for the crystallization and also result in local thinning of the anodic oxide film [27]. The latter is due to poor anodizing of the carbide phase having strong bonds between tantalum and carbon atoms. A strong electrical field in the thinning defects can cause breakdown of the dielectric even at low applied voltage. As an example, Fig. 1.5 shows craters in anodic oxide film enriched with carbon tantalum anode [46].

Oxygen in tantalum and niobium also plays an important role in crystallization of the anodic oxide films on these metals. M. Tierman and R. Millard detected critical oxygen content in tantalum anodes that resulted in a sharp increase in direct current leakage (DCL) in a liquid cell (wet test) [47]. This critical oxygen content was attributed to the saturation of tantalum with oxygen when small oxide crystals precipitate in tantalum anode and speed up the crystallization process. The major contributor to the oxygen content in high CV anodes was a natural surface oxide that dissolved in powder particles during the powder sintering in vacuum.

Critical oxygen content in tantalum anodes associated with the sharp DCL increase in tantalum capacitors was investigated by X-ray diffraction analysis (XRD) [48]. Tantalum atoms form cubic crystalline lattice, while oxygen dissolving in tantalum as interstitial impurity occupies the center of the cube. Increasing oxygen content in tantalum causes expansion of the cubic structure and thereby increases in the period of the crystalline lattice of tantalum. This change in the period of the crystalline lattice can be identified by the XRD analysis by the shift in position of the peaks on the diffraction pattern. As an example, Fig. 1.6 shows (211) diffraction peak for three oxygen contents in tantalum anodes with specific charge 50,000 µC/g and effective primary particle radius $r = 2.1$ µm.

According to Fig. 1.6, (211) diffraction peak is shifting left to the smaller diffraction angles and broadening with increase of oxygen content in tantalum anode. This

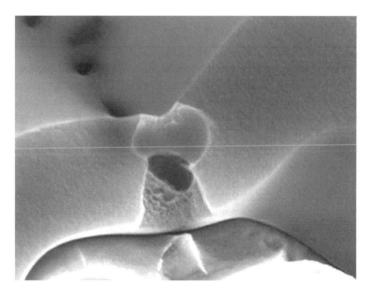

Fig. 1.5 Craters in anodic oxide film enriched with carbon tantalum anode

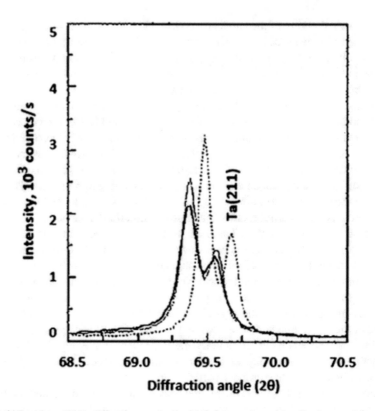

Fig. 1.6 Tantalum (211) diffraction peak obtained for anodes with effective particle radius $r = 2.1$ μm and oxygen content (dotted line) 3000 ppm, (dashed line) 3400 ppm, and (solid line) 3600 ppm

shift in position and shape of the diffraction peak indicates increase in the lattice parameter at higher oxygen contents in tantalum. Lattice parameter reaches its maximum value at the critical oxygen content C_{oc} when DCL increases sharply, while further increase in oxygen content does not affect the lattice parameter (Fig. 1.7).

Flattening lattice parameter with increase of oxygen content above C_{oc} indicates that solid solution of oxygen in tantalum reached solubility limit. Further increase of oxygen in tantalum causes formation of an oxide of tantalum with different crystalline lattices and much higher oxygen content than solubility limit of oxygen in tantalum.

During tantalum powder sintering in vacuum, oxygen from the native oxide can either evaporate into the vacuum chamber or dissolve into the bulk of the tantalum particles. Oxygen evaporates from tantalum at temperatures equal to or above 1880 °C [23, 24] used for sintering of the coarsest tantalum powders with CV/g equal to or below 5000 μC/g. At lower sintering temperatures used for sintering of the higher CV/g tantalum powders, oxygen from the native oxide dissolves into the tantalum particles, increasing the bulk oxygen content in sintered tantalum anodes.

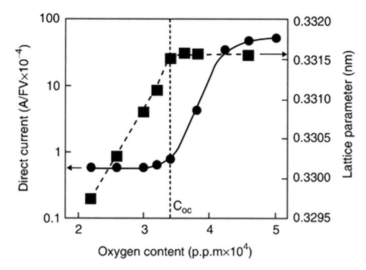

Fig. 1.7 Dependences of the DCL at 50 V formation and lattice parameter versus oxygen content in the tantalum anodes with effective particle radius of 2.1 μm

When the specific charge CV/g of the tantalum powder increases due to decrease in effective particle radius, the ratio between the particle surface and volume becomes larger. This increase of the surface-to-volume ratio results in the larger contribution of oxygen in native surface oxide to the total oxygen content in the tantalum powder and tantalum anodes sintered with this powder. At the same time, lattice parameter of tantalum at critical oxygen content remains practically unchanged with increase of the CV/g of the tantalum powder. This evidences that the major reason of the sharp DCL increase at critical oxygen content is saturation of oxygen content in solid solution of oxygen in tantalum and precipitation of crystalline oxide phase in the bulk of tantalum particles. The XRD analysis allows to distinguish between the total oxygen content in sintered anodes and the bulk oxygen content in the powder particles, which is primarily responsible for the sharp DCL increase at critical oxygen content.

Characteristic cracks from growing crystals can be easily detected on the surface of anodic oxide film on tantalum anodes with critical content (Fig. 1.8) [46].

At the same time, the small precipitates of the crystalline oxide particles appear in the anodic oxide film even when oxygen content in the anode is much below the critical content [46]. In this case, the surface of the anodic oxide film is typically smooth with some scattered bumps and pits reflecting morphology of the underlying tantalum substrate (Fig. 1.9a). After removal of the amorphous film by etching it down to the metal substrate in diluted hydrofluoric acid, one can see peaks of different sizes on the metal surface (Fig. 1.9b).

Electron diffraction analysis confirmed that these peaks are Ta_2O_5 crystals standing on the tantalum surface and hidden in the amorphous matrix before etching. The density and size of these crystalline inclusions on the oxide-metal interface become larger with increasing of oxygen content in the base metal.

Fig. 1.8 Cracks from a
crystal growing in anodic
oxide film on tantalum
anode with critical oxygen
content

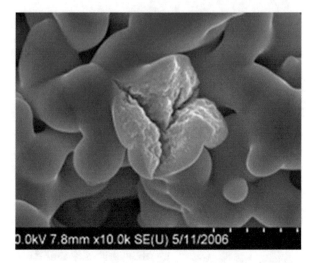

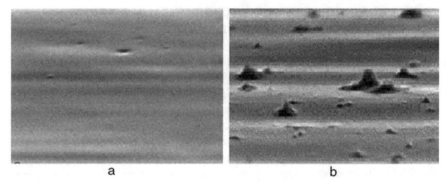

a   b

Fig. 1.9 Top (**a**) and bottom (**b**) of the anodic oxide film on tantalum with 50% of oxygen vs.
saturation

The oxide film, which always exists on the surface of tantalum and niobium
anodes prior to anodizing, can also contribute to the growth of crystalline inclusions
in amorphous matrix of the anodic oxide film [46]. The native surface oxide has
amorphous structure and is only 2–3 nm thick [49]. This native oxide incorporates
into the anodic oxide film during its formation and practically does not influence the
crystallization process. In some cases, low-density initial crystalline inclusions on
tantalum surface with thin native oxide can grow through the whole thickness of the
anodic oxide film without causing mechanical damage. Figure 1.10a shows a crystal
that appears on the surface of the anodic oxide and remains tightly incorporated into
the amorphous matrix. When surface oxide is relatively thick, which is typically due
to the heat release during exothermic passivation process when anodes are exposed
to air after their sintering in vacuum, crystal growth always results in cracking and
disruption of the anodic oxide film (Fig. 1.10b).

Fig. 1.10 Crystallization appearing in the anodic Ta_2O_5 film grown on tantalum with 3 nm (*left*) and 30 nm (*right*) surface oxide

Fig. 1.11 TEM image of the thermal oxide film on tantalum and electron diffraction pattern of the crystalline inclusions in the thermal oxide

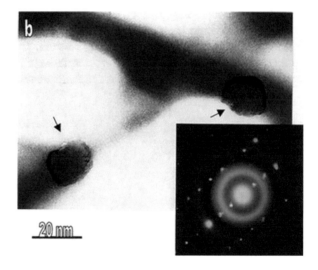

The higher the temperature the anodes were exposed in air prior to anodizing, the thicker is the thermal oxide and the quicker crystallization and disruption of the anodic oxide film appear. The reason for such intense crystallization of the anodic oxide film formed on the relatively thick thermal oxide layer is that the thermal oxide contains small particles of the crystalline phase scattered in the amorphous matrix (Fig. 1.11) [46]. These microcrystals serve as numerous centers for crystallization of the anodic oxide film. The thicker the thermal oxide is, the greater is the portion of the crystalline phase. During anodizing, the thermal oxide incorporates into the growing anodic oxide film forcing up its crystallization.

At given temperature and time, the thickness of the thermal oxide that forms on the surface of tantalum during annealing in air increases with increasing the bulk oxygen content in tantalum [46]. This effect was demonstrated by measuring the oxygen depth profiles annealed in air tantalum foils with different oxygen contents.

Similarly, the thickness of the thermal oxide that forms on the surface of the porous tantalum anodes when they are exposed to air after sintering in vacuum increases with increasing the bulk oxygen content in the tantalum particles. As an example, Fig. 1.12 shows second ion mass spectrometry (SIMS) oxygen depth profiles for the surface of tantalum anodes sintered with 12,000 μC/g tantalum powder and having different bulk oxygen content after the sintering.

As one can see in Fig. 1.12, the thermal oxide is thicker on tantalum anodes with higher bulk oxygen content after sintering, which is indicated by higher and broader SIMS peak of the surface oxide. The formation of the thick thermal oxide layer was caused by oxygen's inability to penetrate the saturated tantalum bulk upon reaching the tantalum surface. The oxygen stayed on the tantalum surface forming the thick thermal oxide layer. The characteristic morphology of the thick thermal oxide was detected by scanning electron microscopy (SEM) on the surface of the tantalum anodes with high bulk oxygen content, while no similar features were detected on the surface of the tantalum anodes with low bulk oxygen content (Fig. 1.13).

Results presented in Figs. 1.12 and 1.13 indicate importance of the procedures used for deoxidizing the tantalum anodes and controlled passivation when the anodes are exposed to air after sintering in vacuum. These processes allow to avoid forming thick thermal oxide on the anode surface and, thereby, suppress field crystallization of the anodic oxide film on tantalum anodes.

Fig. 1.12 SIMS oxygen depth profiles on the sintered tantalum anodes with 12,000 μC/g tantalum powder with different bulk oxygen contents

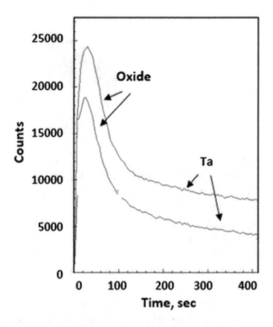

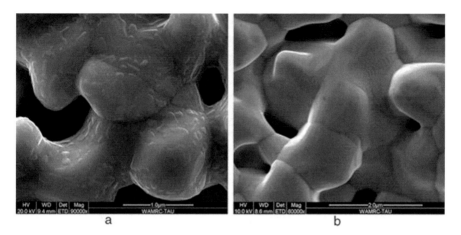

Fig. 1.13 SEM images of the surface of the sintered tantalum anodes with 12,000 μC/g tantalum powder with high (**a**) and low (**b**) bulk oxygen content

1.3   Thermal Crystallization

R. E. Pawel and J. J. Campbell investigated thermal crystallization in anodic oxide films on tantalum and niobium after stripping them from the base metal [50]. They detected that thermal crystallization was initiated in short times above 600 °C for anodic oxide of tantalum and above 500 °C for anodic oxide of niobium. No significant difference was detected in diffraction patterns from crystallized anodic oxide and those from thermal oxide. Many of the crystals observed apparently had not yet grown through the whole thickness of the anodic oxide film. This phenomenon implies that surface nucleation is general to all film thicknesses.

When the Ta-Ta$_2$O$_5$ bilayer was annealed in air, the thermal crystallization was detected at much lower temperatures than that of stripped anodic oxide films [46]. The crystals started to grow during anodizing and continued to grow faster, wider, and higher, further disrupting the anodic oxide film (Fig. 1.14a). The growth was due to the direct access of oxygen from the air to the metal surface through the ruptured anodic oxide film. When annealing was performed in vacuum or inert atmosphere, the crystal growth was much slower than that in air, and the disruption of the anodic oxide film was not followed by the fast buildup of crystal cones (Fig. 1.14b). In this case, re-anodizing after the annealing in vacuum allowed full recovery of the anodic oxide film.

A short-term thermal treatment in vacuum or inert atmosphere can interrupt crystal growth in anodic oxide films [46]. Figure 1.15 shows the bottom of the anodic oxide film on tantalum before (a) and after (b) a 2-min thermal treatment in vacuum at 480 °C (the amorphous matrix was removed by etching in diluted hydrofluoric acid).

One can compare the abundance of small crystals on the tantalum surface before the thermal to the lack of crystals after thermal treatment. Voids and sockets indicate

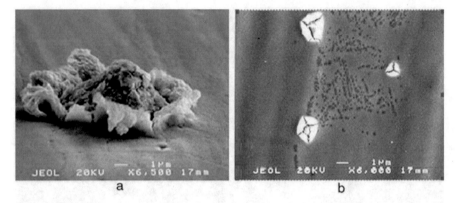

Fig. 1.14 SEM image of the anodic oxide film formed on tantalum foil at 200 V and annealed in air (**a**) or in vacuum (**b**) at 450 °C for 60 min

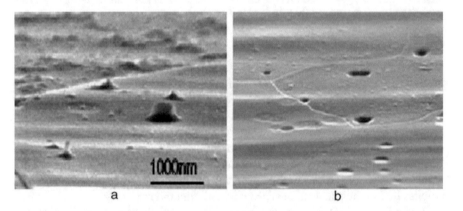

Fig. 1.15 The bottom of the anodic oxide film on tantalum before (**a**) and after (**b**) short thermal treatment in vacuum

the presence of crystalline nuclei before annealing. This means that short thermal treatment caused crystals to detach from the anode surface and withdraw along with the amorphous matrix.

The following presents a physical model for this effect. During anodizing, crystalline nucleus growth takes place at the oxide-metal interface. Consuming tantalum from the anode, the crystals grow down inside the anode as well as up toward the top of the oxide film [28, 41]. This process requires intimate contact between the crystalline lattice of the oxide and the crystalline lattice of the base metal. During thermal treatment, the amorphous oxide film undergoes high mechanical stress due to the large difference in the coefficient of the thermal expansion of the metal and oxide film [51]. This mechanical stress causes the oxide film to slide (shear) along the oxide-metal interface and cut the crystals sitting inside the metal. Losing intimate contact with the anode, the crystals bury in the amorphous matrix and lose the ability for further growth.

1.4   Interaction Between Oxygen Migration and Crystallization

Oxygen migration through the anodic oxide film and its interfaces with the anode and cathode in tantalum- and niobium-based capacitors has a strong influence on the crystallization process in anodic oxide films on tantalum and niobium. G. Klein demonstrated that extraction of oxygen from the anodic oxide film by the base metal proceeds in a nonhomogeneous way with a higher activity in the vicinity of crystalline grain boundaries, dislocations, etc. [52]. After the thermal treatment, anodic oxide film in these areas was characterized by strong depletion of oxygen, and the adjacent metal layer was strongly enriched with oxygen.

Structural sequences of this local oxygen redistribution were different for anodic oxide films on tantalum and on niobium [19]. In anodic oxide film on tantalum with steep oxygen gradient, crystalline inclusions of tantalum pentoxide (cr-Ta_2O_5) are formed in amorphous matrix of the film (a-Ta_2O_5) at the interface with the base metal (Fig. 1.16 left). In anodic oxide film on niobium, oxygen-depleted zones expanded through the whole oxide film thickness due to a higher oxygen diffusion coefficient in Nb_2O_5 than that in Ta_2O_5. This resulted in the formation of non-stoichiometric channels (a-Nb_2O_{5-x}) with increased electrical conductivity in amorphous matrix of the anodic oxide film on niobium (Fig. 1.16 right).

Presented in Fig. 1.16 was a model developed to explain different effects of thermal treatment on current-voltage $I(V)$ characteristics of anodized tantalum and niobium foils [19]. A clipping probe made of the same material as the base metal was

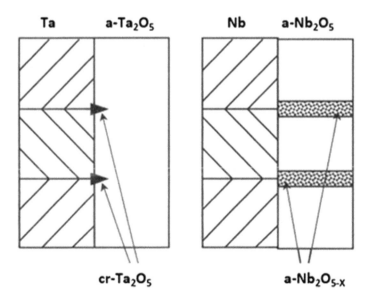

Fig. 1.16 Scheme of the transformations in Ta_2O_5 and Nb_2O_5 caused by the thermal treatment of metal-oxide bilayers

used as a counter electrode. Figure 1.17 presents $I(V)$ characteristics of Ta-Ta$_2$O$_5$-Ta and Nb-Nb$_2$O$_5$-Nb capacitors before and after 1 h thermal treatment at 350 °C for anodized tantalum foil and 300 °C for anodized niobium foil. The $I(V)$ characteristics were measured at positive and negative polarity of the base metal.

According to Fig. 1.17, $I(V)$ characteristics of the Ta-Ta$_2$O$_5$-Ta and Nb-Nb$_2$O$_5$-Nb capacitors at positive and negative polarity of the base electrode were similar to each other with little or no asymmetry before the thermal treatment. After the thermal treatment, $I(V)$ characteristics of tantalum capacitor became strongly asymmetric due to the significant shift of the negative branch to higher currents with a little change in the positive branch. In contrast, $I(V)$ characteristics of niobium capacitor remained symmetric after the thermal treatment due to the shift of both positive and negative branches to higher current magnitudes.

$I(V)$ characteristics of the thermally treated tantalum capacitor can be explained by the inability of crystalline inclusions in anodic oxide film on tantalum to affect the current at positive polarity of the base electrode (forward voltage) while they are tightly incorporated in the amorphous matrix and have a small size with regard to the total oxide film thickness. Only when the crystals reach a critical size and disrupt the anodic oxide film does the current increase sharply. At the same time, the current in niobium capacitor increases gradually with an increase in the diameter and density of the conductive channels in the anodic oxide film.

Hidden in the anodic oxide film on tantalum, crystalline inclusions don't show themselves in the current at forward voltage but may be displayed at reverse voltage. When negative polarity is applied to the tantalum base electrode, these crystalline inclusions work as powerful injectors of electrons due to the concentration of the electrical field on the spikes pointed from the base electrode (cathode at negative polarity) to the dielectric. The injection of electrons allows elevated current at

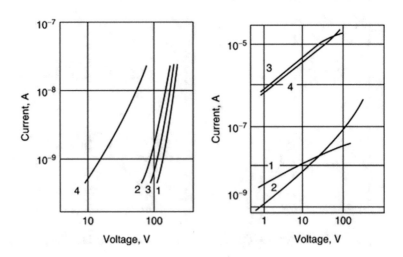

Fig. 1.17 $I(V)$ characteristics of Ta-Ta$_2$O$_5$-Ta (*left*) and Nb-Nb$_2$O$_5$-Nb (*right*) capacitors before (1,2) and after (3,4) thermal treatment (1,3—positive polarity, 2,4—negative polarity)

reverse voltage to indicate the initial stage of the formation of the crystalline inclusions in anodic oxide film of tantalum [19].

As presented in Fig. 1.16, transformations in Ta-Ta$_2$O$_5$ and Nb-Nb$_2$O$_5$ bilayers as a result of their thermal treatment may also explain different types of the DCL failures in Solid Electrolytic Tantalum and niobium capacitors, which undergo multiple thermal treatments during pyrolytic deposition of the MnO$_2$ cathode. Typically, the DCL in a "bad" tantalum capacitor stays stable or even decreases with time during long-term testing or in the field and then jumps up sharply, resulting in catastrophic failure of tantalum capacitor. In contrast, the DCL in a "bad" niobium capacitor increases gradually with time, resulting in parametric failure of the capacitor [53]. Small initial crystalline inclusions in the dielectric of tantalum capacitor don't affect the DCL until they reach the critical size and disrupt the dielectric, causing the sharp DCL increase. At the same time, the DCL in niobium capacitor increases gradually due to the increase in the diameter and density of the non-stoichiometric conductive channels in the dielectric.

As shown in [9, 19], oxygen migration at any given temperature is more active in niobium capacitors than in tantalum capacitors. Due to active oxygen migration, niobium anodes practically saturate with oxygen during high-temperature processing related to the pyrolytic deposition of the MnO$_2$ cathode (Fig. 1.18) [54].

No additional oxygen redistribution or phase transformations were detected in niobium anodes after a 1000 h life test at a rated voltage and 85 °C. This was expected because the anodes have been saturated with oxygen prior to the test. At the same time, a small area of crystalline anodic oxide film was detected in the dielectric of a capacitor that failed short with the test and then was stripped from external layers and the MnO$_2$ cathode (Fig. 1.19a). The anodic oxide film had an amorphous structure outside of this crystalline area (Fig. 1.19b).

These results show that crystallization of the amorphous matrix of the dielectric also contributes to the failures of niobium capacitors during long-term testing or in the field, especially when niobium anodes become saturated with oxygen during pyrolytic deposition of the MnO$_2$ cathode.

During sintering of tantalum or niobium powder, oxygen diffuses from oxygen-enriched powder particles into the relatively pure tantalum or niobium wire embedded into the pellet during the powder pressing. Depending on the powder sintering temperature, the length of oxygen diffusion into the wire varies from 0.1 mm to 1 mm. In this area, oxygen preferably locates in crystalline grain boundaries of the wire rather than in the crystal bulk. When local oxygen content in the crystalline grain boundaries reaches the solubility limit, crystals of the oxide phase precipitate from the solid solution of oxygen in tantalum or niobium (Fig. 1.20) [55].

The harder oxide particles push the metal crystals apart, resulting in the oxygen-induced wire brittleness. Increased sintering temperature causes crystalline grains in the wire to become larger and shortens the crystalline grain boundaries. This allows for easier local oversaturation of the wire with oxygen and precipitation of the oxide phase, especially in the thinner wires. Only at the very high sintering temperatures used for the coarsest powders that oxygen evaporates in vacuum out of the tantalum and niobium anodes, eliminating the oxygen-induced wire brittleness.

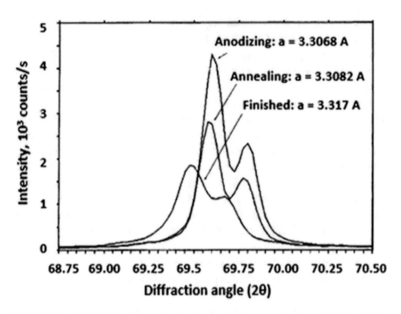

Fig. 1.18 Niobium (211) diffraction peak obtained in niobium anode after anodizing, first annealing during deposition of MnO_2 cathode, and in finished niobium capacitor

Fig. 1.19 Crystalline (**a**) and amorphous (**b**) areas in anodic oxide film in niobium capacitor that failed short at life test

Hydrogen can also cause brittleness of the lead wire embedded in tantalum and niobium anodes due to the transformation of the microstructure, which establishes local high concentration of hydrogen and mechanical stress [56]. Typically hydrogen-induced brittleness occurs when the anodes are exposed to humid air or submerged in aqueous solutions after their sintering in vacuum, provoking surface oxide growth and releasing hydrogen. As it was demonstrated in Figs. 1.12 and 1.13, surface oxidation and, thereby, hydrogen pickup intensify with increasing the

Fig. 1.20 Precipitates of oxide phase in crystalline grain boundaries in tantalum wire

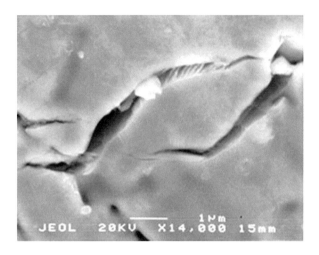

bulk oxygen content in the sintered anodes, making hydrogen-induced wire brittleness more pronounced in the anodes with higher CV/g tantalum and niobium powder. Due to the high mobility of hydrogen in metals even at room temperature, hydrogen-induced brittleness spreads through the entire length of the lead wire. Special ways of passivation of the sintered tantalum and niobium anodes help mitigate brittleness of the lead wire [57, 58].

In conclusion, the two major mechanisms of degradation of tantalum- and niobium-based capacitors are oxygen migration from the anodic oxide film of tantalum and niobium to the base metal (Ta, Nb, NbO) and crystallization of the amorphous matrix of the anodic oxide film. Both mechanisms originate from the thermodynamic instability of the anodic oxide film and its interface with the anode. Although both degradation mechanisms work simultaneously, one of them may have a dominant effect on the capacitor performance and reliability. In general, low-voltage capacitors are more susceptible to oxygen migration, while the high-voltage capacitors are more susceptible to crystallization. Despite the thermodynamic nature of these degradation mechanisms, the pace of the degradation processes can be reduced to a negligible level by technological means, allowing manufacturing of tantalum- and niobium-based capacitors with exceptional stability and reliability during any practical duration of the testing and field application.

Chapter 2
Basic Technology

The manufacturing of tantalum and niobium-based capacitors begins with the pressing and sintering of tantalum, niobium, or NbO anodes and ends with the testing and packaging of the finished capacitors. The major stages of the manufacturing are the making of porous anodes, growing of either Ta_2O_5 or Nb_2O_5 dielectric on the surface of porous anodes, forming of either liquid electrolyte, MnO_2, or conductive polymer cathode, coating with external layers of carbon and silver in solid capacitors, and testing and packaging of finished capacitors.

2.1 Anode Manufacturing

Physical properties and chemical composition of the capacitor-grade powders play a critical role in the performance and reliability of the capacitors made with this powder. The first capacitor-grade tantalum powders were manufactured by electron-beam purification of the tantalum ingot, hydration and crushing of the ingot, milling of fragments into powder, and dehydration of the powder by thermal treatment in vacuum. The resulting powder with specific charge equal to or below 10,000 µC/g, usually called electron-beam or EB powder, has coarse primary particles with irregular shape and an average size of several microns (Fig. 2.1).

EB tantalum powders are typically pressed at 8–9.5 g/cm³ density and sintered in vacuum at temperatures 1800–2100 °C. High sintering temperatures provide further purification of tantalum anodes of most metallic and nonmetallic impurities. A combination of high press density and high sintering temperature allows for the growth of thick necks between powder particles that are important for anodizing at the high voltages typically associated with this powder. The EB tantalum powders are still used in the manufacturing of the highest-voltage tantalum capacitors due to the large-size particles, large necks between the sintered particles, and high chemical purity.

© The Author(s), under exclusive license to Springer Nature Switzerland AG 2022   23
Y. Freeman, *Tantalum and Niobium-Based Capacitors*,
https://doi.org/10.1007/978-3-030-89514-3_2

Fig. 2.1 SEM image of EB tantalum powder with specific charge 3400 µC/g

The higher CV/g tantalum powders are typically manufactured by sodium reduction of potassium tantalum fluoride (K_2TaF_7). Th process performs in liquid phase at temperatures around 600 °C with constant agitation of the liquid inside the reactor. The following chemical reaction takes place:

$$K_2TaF_7 + 5Na \rightarrow Ta + 2KF + 5NaF$$

During sodium reduction, heavier tantalum particles drift to the bottom of the reactor, while lighter inert salts of potassium and sodium fluorides drift close to the surface. The physical properties of the sodium-reduced tantalum powders such as average size of the primary particles and specific surface area are controlled by inert salts such as KCl, KF, NaCl, and NaF that are added to the reduction mixture [59]. As the proportion between the inert salts and K_2TaF_7 increases, the resulting tantalum powder becomes finer with smaller primary particles and higher specific charge. As an example, Fig. 2.2 shows SEM images of tantalum powders with different specific charges obtained by sodium reduction of potassium tantalum fluoride.

Sodium-reduced tantalum powders undergo water washing and leaching in mineral acids such as HCl and HNO_3 to dissolve and remove inert salts. The following steps of vacuum heat treatment, milling, deoxidizing by magnesium vapor, leaching of magnesium oxide in diluted sulfuric acid, and final water washing and drying improve flow ability and chemical purity of the tantalum powder. Some powders are treated with dopants such as phosphorus or boron, which act as sintering retardants and allow for higher sintering temperatures without significant shrinkage and loss of surface area.

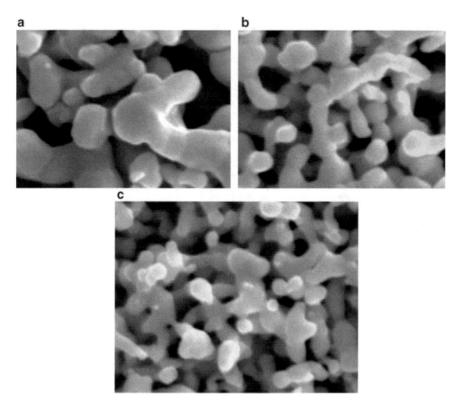

Fig. 2.2 SEM images of sodium-reduced tantalum powders with specific charge 50,000 µC/g (**a**), 100,000 µC/g (**b**), and 150,000 µC/g (**c**) (×10,000)

Alkali metals such as potassium and sodium, used in sodium reduction of potassium tantalum fluoride, have low melting temperatures and cannot be dissolved in the bulk of tantalum particles. Residuals of these metals on the surface of tantalum particles are evaporated during vacuum heat treatment following sodium reduction process. However, complex alkali-tantalite compounds with high melting temperatures such as $KTaO_3$ and $NaTaO_3$ can form during sodium reduction of K_2TaF_7 when oxygen is present in the reactor volume [60]. Residuals of moisture in the K_2TaF_7 precursor can be a source of oxygen in the reactor. Due to their high melting temperature and chemical stability, small precipitates of the alkali-tantalite compounds can remain in tantalum powder and sintered anodes and then be incorporated into the anodic oxide film during its formation. As an example, Fig. 2.3 shows breakages of tantalum anodes enriched with potassium and pure tantalum anodes formed to 250 V. Bright areas on these images represent the core of the tantalum particles, and gray areas around tantalum core represent anodic oxide on tantalum.

As one can see in Fig. 2.3, there are numerous voids inside the anodic oxide formed on tantalum enriched with potassium, while there are no defects inside the anodic oxide formed on pure tantalum. These voids are associated with small

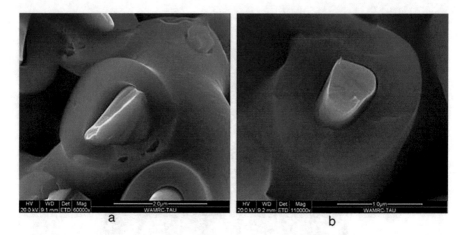

a   b

Fig. 2.3 SEM images of the breakages of tantalum anodes enriched with potassium (**a**) and pure tantalum anodes (**b**) formed to 250 V

precipitates of potassium-tantalite into the tantalum particles, which were incorporated into the anodic oxide film during its formation. Tantalum capacitors with these defects in the dielectric have higher DC leakage (DCL) and lower breakdown voltage (BDV) in comparison to DCL and BDV in tantalum capacitors with defect-free dielectrics. These results show critical importance of drying the K_2TaF_7 precursor and removal from the reactor other sources of oxygen in the sodium reduction process.

Some tantalum powders are produced by direct reduction of tantalum pentoxide powder by magnesium vapor [61, 62]. The following exothermic reaction takes place in this case:

$$Ta_2O_5 + 5Mg \rightarrow 2Ta + 5MgO$$

Unlike the EB and sodium reduction processes, the direct magnesium reduction process does not require the melting of Ta_2O_5 precursor. The morphology, chemical composition, and physical properties of the solid Ta_2O_5 precursor can be used to produce tantalum powders with a broad range of specific surface area and charge. Typically, high CV/g tantalum powders with specific charge equal to or above 100,000 µC/g are produced by direct magnesium reduction. As an example, Fig. 2.4 shows a SEM image of the tantalum powder with specific charge 200,000 µC/g and surface area 4.2 m²/g obtained by direct magnesium vapor reduction of Ta_2O_5 powder.

Every batch of finished tantalum powder produced by either EB, sodium reduction, or direct magnesium reduction processes is provided with a certificate of quality, which shows actual chemical, physical, and electrical properties of the powder. Chemical composition includes alkali and transition metals, carbon, oxygen, and other impurities. The physical properties include average size of the primary particles, bulk density without pressing (Scott density), flowability, size distribution of

Fig. 2.4 SEM image of 200,000 μC/g tantalum powder obtained by direct magnesium vapor reduction of Ta_2O_5 powder

the agglomerated particles, etc. The electrical properties of the powder include specific charge (CV/g), direct current leakage per unit of specific charge (DCL/CV), equivalent series resistance (ESR), and breakdown voltage (BDV) for a given formation voltage. Pressing, sintering, and formation conditions used to make and test electrical properties of the powder are also included in the certificate of quality.

Anode manufacture typically begins with mixing of tantalum powder with organic lubricant to increase the powder flowability, improve density uniformity and crush strength in the pressed anodes, and reduce wear-out of the dies and punches. A small percentage of tantalum anodes are manufactured "green" without adding organic lubricant to avoid contamination with carbon from the organic lubricant. The EB powders with larger primary particles are typically pressed with 8–9 g/cm^3 press density, while sodium-reduced and direct magnesium-reduced tantalum powders with finer primary particles are typically pressed with 5–6 g/cm^3 press density. Tantalum lead wire is typically embedded into the anode during its pressing as a positive termination of the tantalum capacitor; however, in some cases the lead wire is welded to the pre-sintered anodes to improve the powder-wire bond. Pressed tantalum anodes have different shapes and sizes depending on the type of the finished tantalum capacitors utilizing these anodes (Fig. 2.5).

Shown in Fig. 2.5, cylindrical anodes are used in hermetic tantalum capacitors with through-hole assembly, while rectangular anodes are typically used in

Fig. 2.5 Pressed tantalum anodes

surface-mount tantalum capacitors with pick-and-place assembly. Some tantalum anodes are pressed fluted to increase their external surface area and, thereby, decrease the ESR of the tantalum capacitor [63, 64].

After the pressing, the organic lubricant is decomposed and removed from the tantalum anodes by vacuum heat treatment in the de-lubrication furnace at temperatures 200–300 °C depending on the type of the lubricant. After the de-lubrication, tantalum anodes are placed in the sintering furnace where they are sintered in vacuum at different temperatures depending on the specific charge of the powder. During sintering, tantalum anodes shrink, while the necks between the powder particles grow, providing mechanical strength to the sintered anodes. The moving force for the sintering is reducing surface area and, thereby, internal energy of the system of the powder particles. This force is inversely proportional to the average size of the powder particles, resulting in more rapid shrinkage of the finer powders with smaller primary particles [65]. To avoid excessive shrinkage and loss of surface area, the finest tantalum powders with submicron-sized primary particles are sintered at about 1200 °C. Sintering temperature gradually increases for the coarser tantalum powders reaching about 2100 °C for the coarsest EB tantalum powders.

Press density and sintering temperature of a given tantalum powder determine the specific charge CV/g of the sintered tantalum anodes. As an example, Fig. 2.6 shows CV/g as a function of press density (d) and sintering temperature (T_s) for tantalum anodes sintered with tantalum powder with an average primary particle size of 3.7 μm [66].

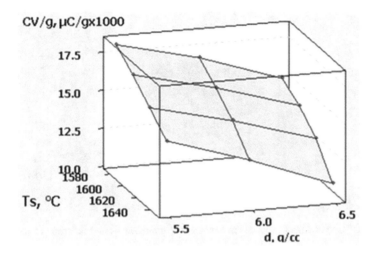

Fig. 2.6 *CV*/g of tantalum anodes as a function of press density and sintering temperature

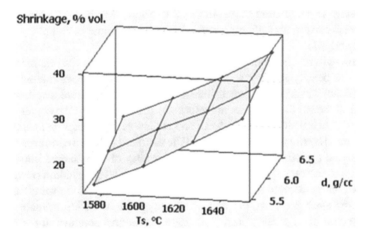

Fig. 2.7 Shrinkage of tantalum anodes as a function of press density and sintering temperature

According to Fig. 2.6, *CV*/g decreases with both press density and sintering temperature. This *CV*/g decrease is due to the shrinkage of the tantalum anodes and related loss in surface area. Figure 2.7 shows dependence of the shrinkage on press density and sintering temperature for the same powder as in Fig. 2.7.

As one can see in Figs. 2.6 and 2.7, decrease in *CV*/g with higher press density and sintering temperature is "mirrored" by corresponding increase in shrinkage with these parameters. Tantalum anodes typically shrink about 4–6% during sintering, providing sufficient mechanical strength to the porous anodes and strong bonds between the powder particles and embedded lead wire.

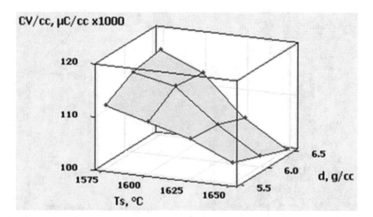

Fig. 2.8 CV/cm^3 of tantalum anodes as a function of press density and sintering temperature

The dependence of volumetric charge efficiency (CV/cm^3) on press density and sintering temperature is different from the dependence of weight charge efficiency (CV/g) on these parameters shown in Fig. 2.6. Figure 2.8 shows CV/cm^3 as a function of press density and sintering temperature for the same tantalum powder as in Figs. 2.6 and 2.7.

According to Fig. 2.8, maximum CV/cm^3 can be achieved at a combination of the higher press density and lower sintering temperature. This combination also provides sufficient shrinkage and mechanical strength to the sintered tantalum anodes.

A critical part of the anode manufacturing is the exposure of the tantalum anodes to air after they are sintered in vacuum and cooled down to room temperature. The heat generated by the surface oxidation of the tantalum particles can cause excessive thermal oxide growth or even igniting and burning of the sintered anodes. Most susceptible to the rapid temperature increase are high CV/g tantalum powders with a high surface-to-volume ratio in the primary particles. To minimize effects of the rapid temperature increase, the rate of the oxygen flow inside vacuum chamber should be controlled to allow heat dissipation via thermal conductivity of the tantalum anodes and tantalum crucibles with these anodes. When the temperature stays close to room temperature during exposure of the tantalum anodes to air after sintering, only a thin native oxide with amorphous structure forms on the surface of the tantalum particles. This native oxide is incorporated into the anodic oxide film on tantalum during anodizing process without any negative effect on the performance and reliability of tantalum capacitors.

2.2   Ta$_2$O$_5$ Dielectric

After the sintering, tantalum anodes are welded by the lead wire to the carrier bars, typically made of aluminum or stainless steel, and proceed though the following steps of the manufacturing of tantalum capacitors starting with formation of the

Ta$_2$O$_5$ dielectric. As an example, Fig. 2.9 shows rectangular tantalum anodes welded to the stainless steel carrier bar.

To form the Ta$_2$O$_5$ dielectric, the carrier bars with tantalum anodes are suspended in an electrolyte solution and anodized under appropriate electrical conditions. During anodizing, positive potential from the power supply is applied to the carrier bar, and negative potential is applied to the formation tank. The anodizing electrolyte typically consists of a dilute aqueous solution of a mineral acid such as phosphoric, sulfuric, or nitric acid. The electrolytes providing the best dielectric quality to the anodic oxide film on tantalum contain water, phosphoric acid, and at least one organic solvent such as polyethylene glycol dimethyl ether [67, 68].

Formation of anodic oxide film on tantalum anodes is typically performed in two stages: the initial constant current stage in which voltage is linearly increasing with time until achieving a set formation voltage followed by the constant voltage stage in which current is decaying with time [17, 67–70]. As an example, Fig. 2.10 demonstrates current and voltage dependence on time for the two-stage formation of tantalum anodes (provided by J. T. Kinard).

The amount of anodic oxide film formed on tantalum surface during formation is directly proportional to the current passed through the electrolytic cell, which results in linear increase of voltage with time during the constant current stage of formation. At this stage of formation, the current has an ionic nature with very little contribution of the electronic current. Under a strong electric field in the dielectric, negatively charged ions of oxygen diffuse from the liquid electrolyte toward the interface with tantalum anode, and positively charged ions of tantalum diffuse from the anode toward the interface with the liquid electrolyte. The redox reactions on these interfaces provide the dielectric growth. At constant voltage stage of formation, current decays with time due to the completion of the dielectric growth and

Fig. 2.9 Tantalum anodes welded to the carrier bar

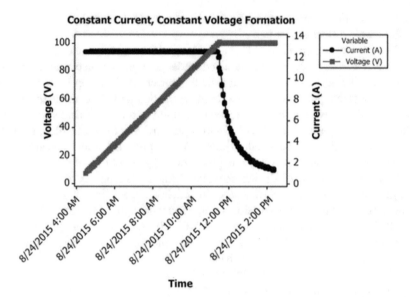

Fig. 2.10 Current and voltage dependence on time for two-stage formation of tantalum anodes

achieving the stoichiometric composition through the whole thickness of the dielectric. At the end of formation, there is only small electronic DC leakage current running through the dielectric at formation voltage [17].

At constant electrolyte temperature, the thickness of the anodic oxide film on tantalum is directly proportional to the formation voltage with the coefficients 1.63 nm/V at room temperature and about 2 nm/V at 80 °C [17]. At constant formation voltage, the thickness of the anodic oxide film is directly proportional to the Kelvin (absolute) temperature of the electrolyte. This relationship between the thickness of the dielectric and absolute temperature provides the simple relationship between the capacitance C and absolute temperature T of the formation electrolyte at constant formation voltage:

$$C_1 T_1 = C_2 T_2,$$

where C_1 is capacitance at temperature T_1 and C_2 is capacitance at temperature T_2. This relationship was found to work over a range of 5–500 V and 0–200 °C [71].

The major parameters of the formation process are formation current at constant current stage of formation and formation voltage and hold time at constant voltage stage of formation. The formation current is typically in the range of 0.5–5 A/C, decreasing in anodes with finer pores sintered with high CV/g tantalum powders. Slower heat exchange between the electrolyte inside and outside of these anodes, especially in larger size anodes, can overheat and damage the dielectric. The formation voltage is typically 1.8–2.5 times higher than rated voltage in Wet Tantalum capacitors and 2.5–4 times higher than rated voltage in Solid Electrolytic and

Polymer Tantalum capacitors. These ratios between the formation voltage and rated voltage provide long-term stability and reliability to tantalum capacitors. The hold time at formation voltage should be long enough to allow the current to decay and flatten, but not too long to provoke field crystallization and, thereby, current increase.

An additional stage of constant power can be added to the formation process in order to suppress overheating and field crystallization [67]. As an example, Fig. 2.11 demonstrates current and voltage dependence on time for three-stage formation of tantalum anodes (provided by J. T. Kinard).

As one can see in Fig. 2.11, the formation begins with the constant current stage when voltage increases linearly with time and then continues with the constant power stage when voltage increases at a lower rate, while current gradually decreases with time and finally changes to the constant voltage stage when set formation voltage is achieved and current decays with time. In combination with low-temperature formation in an electrolyte containing polyethylene glycol dimethyl ether, the three-stage formation allows for the highest formation voltages in sintered tantalum anodes [68].

After the formation of the Ta$_2$O$_5$ dielectric is completed, tantalum anodes are rinsed in DI water to remove residuals of the formation electrolyte. The color of the formed tantalum anodes has an interferential nature and depends on the thickness of the anodic oxide film, which is defined by formation voltage and temperature of the formation electrolyte. As an example, Fig. 2.12 shows tantalum anodes formed to 105 V at 80 °C to achieve the dielectric thickness 210 nm.

After the initial formation, tantalum anodes are typically subjected to the heat treatment in the range of temperatures 250–450 °C following by final formation. As

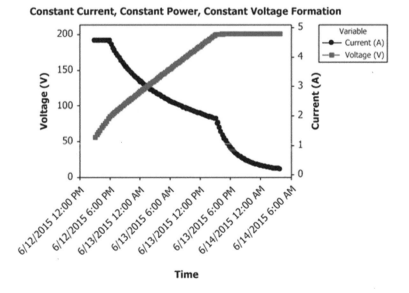

Fig. 2.11 Current and voltage dependence on time for three-stage formation of tantalum anodes

Fig. 2.12   Tantalum anodes formed to 105 V at 80 °C

it was described earlier, during the heat treatment, oxygen migrates from the anodic oxide into the tantalum anode, leaving oxygen vacancies in the oxide film. The final formation, typically at the same voltage and temperature as these at the initial formation, restores stoichiometric composition of the anodic oxide of tantalum. At the same time, tantalum anode near its interface with the anodic oxide remains enriched with oxygen, suppressing further oxygen migration and stabilizing chemical composition of the dielectric. The heat treatment is typically performed in air except when tantalum anodes are sintered with fine high CV/g tantalum powder and formed to low voltages. In this case the heat treatment is performed in vacuum or inert atmosphere to prevent possible thermal oxide formation under the thin anodic oxide [25] or even ignition of tantalum anodes. Similar to shown in Fig. 1.13a thermal oxide grows on the surface of tantalum with high bulk oxygen content typical for high CV/g tantalum anodes. The heat treatment improves DC leakage stability during manufacturing, testing, and field application of the finished tantalum capacitors. The heat treatment in vacuum or inert atmosphere also improves dissipation factor (DF) regardless of the anode CV/g and the dielectric thickness. Some increase in capacitance can be achieved by the heat treatment in air because of pre-ordering of the amorphous matrix of the dielectric and, thereby, dielectric constant increase [49].

   In some cases, shell formation is performed before the final formation to grow a thicker protective layer of the dielectric around the external anode surface, where there is a higher probability of damage to the dielectric during the capacitor manufacturing and testing. The shell formation is performed in aqueous electrolytes containing alkali metal salt of organic acid using a rapid voltage increase above the initial formation voltage and short hold time at the shell formation voltage [72]. Typically shell formation causes little loss in the capacitance since it affects only very thin external layer of the tantalum anodes.

2.3   Cathode

The most significant breakthroughs in performance of tantalum capacitors were achieved due to changes in material of their cathode. The majority of currently manufactured tantalum capacitors are solid with either MnO_2 or conductive polymer cathodes. At the same time, tantalum capacitors with liquid electrolyte inside a hermetic case are still used in modern electronics due to their high reliability, working voltage, and volumetric efficiency. The internal surface of the hermetic case is employed as the cathode in Wet Tantalum capacitors. In these capacitors, the anode capacitance C_a and the cathode capacitance C_c are connected to each other in series by liquid electrolyte, providing a total capacitance C equal to

$$\frac{1}{C} = \frac{1}{C_a} + \frac{1}{C_c}$$

$$C = \frac{C_a \times C_c}{C_a + C_c}$$

From these equations, the capacitance of Wet Tantalum capacitors approaches the anode capacitance, $C \approx C_a$, when the cathode capacitance is much larger than the anode capacitance, $C_c \gg C_a$. To increase cathode capacitance, the inside wall of the case has a porous coating made with oxide of ruthenium or several other metals [73–75]. Other types of Wet Tantalum capacitors have a porous sleeve made of fine tantalum powder and sintered to the inside wall of the tantalum case. The sleeve has a thin anodic oxide film that improves reverse voltage and ripple current characteristics of the Wet Tantalum capacitors [75].

In Solid Electrolytic Tantalum capacitors, the deposition of the MnO_2 cathode on the surface of the Ta_2O_5 dielectric follows the original US patent on these capacitors [5]. Formed tantalum anodes are immersed in an aqueous solution of manganese nitrite and then transferred to an oven where the manganese nitrite is converted into manganese dioxide by pyrolytic reaction:

$$Mn(NO_3)_2 \rightarrow MnO_2 + 2NO_2 \uparrow$$

The conversion of manganese nitrite into manganese dioxide is typically performed in the temperature range of 260–320 °C. Higher temperatures activate pyrolysis of manganese nitrite, but also intensify oxygen migration from the underlying anodic oxide film into the tantalum anode and increase the probability of thermal crystallization of the film's amorphous matrix. Actual temperature of pyrolytic conversion of the manganese nitrite into manganese dioxide is chosen as a balance between these positive and negative trends.

The steps of immersion of the formed tantalum anodes into the aqueous solution of manganese nitrite and its pyrolytic conversion into manganese dioxide are repeated several times with a gradual increase of the viscosity of aqueous manganese nitrite solution. These repetitions insure full coverage of the Ta_2O_5 dielectric with the MnO_2 cathode inside and outside of the porous tantalum anodes. The pyrolytic conversion of manganese nitrite into manganese dioxide is usually performed in a water steam atmosphere, which increases density and, thereby, conductivity of the MnO_2 cathode providing lower equivalent series resistance in the finished capacitors. After several steps of impregnation with the MnO_2 cathode, formed tantalum anodes undergo reformation in liquid electrolyte to restore stoichiometric composition of the Ta_2O_5 dielectric. The reformations help reduce DC leakage and improve the stability and reliability of Solid Electrolytic Tantalum capacitors during long-term testing and field application.

In Polymer Tantalum capacitors, highly conductive and thermally stable conjugated polymer poly(3,4-ethylenedioxythiophene) (PEDOT) is commonly used as the cathode material [14–16]. A schematic of PEDOT is shown in Fig. 2.13 [76].

In Polymer Tantalum capacitors, the deposition of the PEDOT cathode on the surface of the Ta_2O_5 dielectric is typically performed by either in situ chemical oxidation polymerization or by the use of pre-polymerized PEDOT suspension. In situ oxidative polymerization of PEDOT is performed by chemical reaction between monomer 3,4-ethylenedioxythiophene and oxidant solution of iron (III) p-toluenesulfonate (FePTS) with a monomer/oxidant ratio of 3:1 [16, 77, 78].

The PEDOT cathode obtained by in situ chemical reaction is a mixture of polymer and paratoluene sulfonic acid (pTSA), which works as a dopant increasing conductivity of the PEDOT cathode (Fig. 2.14).

Pre-polymerized PEDOT is applied by dipping the formed tantalum anodes into a waterborne dispersion of nanoscale PEDOT particles and the dopant poly(styrene sulfonate) (PSS). Anodes are subsequently dried in air at room temperature and then at about 150 °C [79, 80]. The schematic of the pre-polymerized PEDOT/PSS is shown in Fig. 2.15.

The formed tantalum anodes are perpendicularly dipped and withdrawn into the PEDOT dispersion surface with a rate of 0.1–0.5 mm/s and a hold time of 30–300 s. In both in situ and pre-polymerized processes, 5–6 cycles of PEDOT application provide maximum coverage of the Ta_2O_5 dielectric with the conductive polymer cathode inside and outside of the porous tantalum anodes [81].

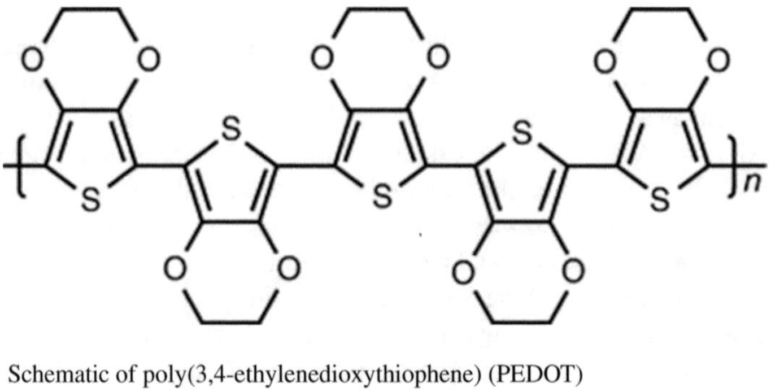

Fig. 2.13 Schematic of poly(3,4-ethylenedioxythiophene) (PEDOT)

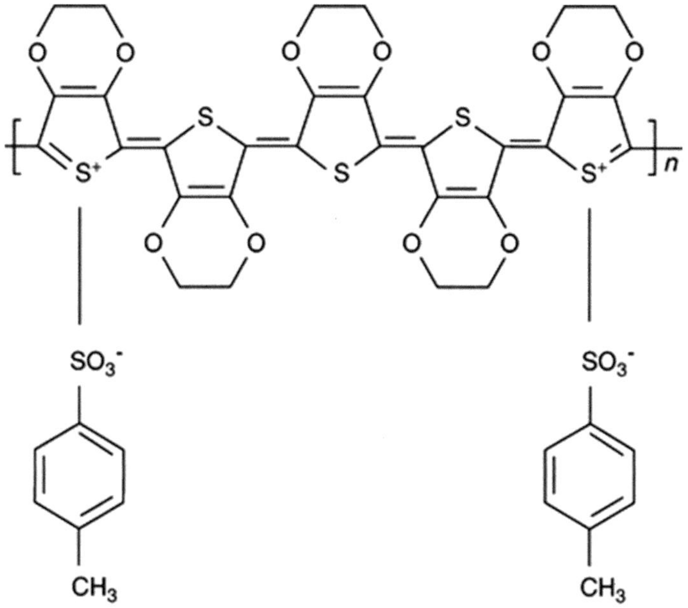

Fig. 2.14 Schematic of PEDOT with pTSA as a dopant

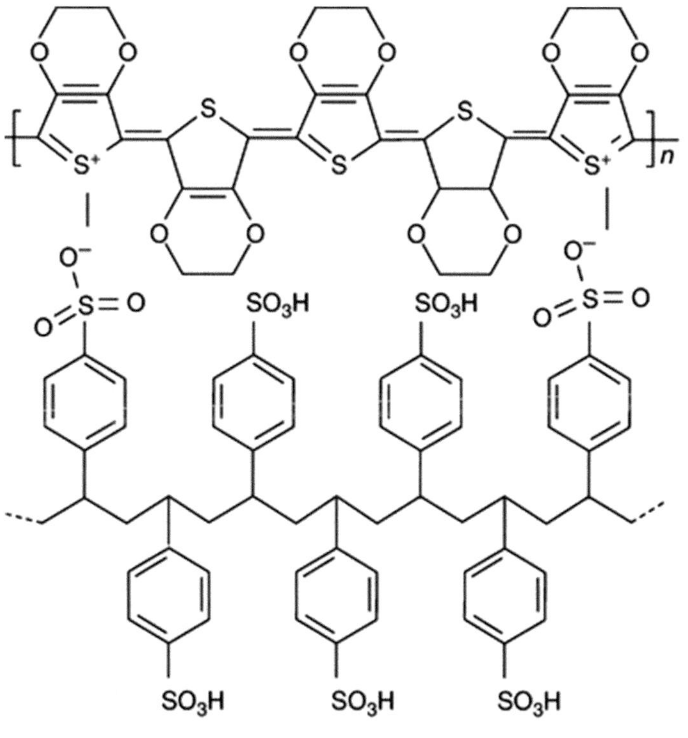

Fig. 2.15 Schematic of pre-polymerized PEDOT with PSS as a dopant

2.4   External Layers, Encapsulating, and Testing

The external layers of carbon and silver are applied on top of the capacitor elements with either MnO_2 or conductive polymer cathode. Silver reduces ESR and provides negative termination to the solid tantalum capacitors. Carbon prevents the migration of silver from the external silver layer toward the Ta_2O_5 dielectric, which is especially active at higher temperatures and ambient humidity and causes catastrophic failures of solid tantalum capacitor. In tantalum capacitors with MnO_2 cathode, the carbon layer also prevents oxidation of the external silver layer by the MnO_2 cathode that causes ESR increase during the capacitor testing and field application.

The external carbon layer is formed by dipping capacitor elements into a water-based or organic solvent-based dispersion of carbon black or graphite particles and organic binder. After air-drying the capacitor elements undergo heat treatment at temperatures about 150 °C, which evaporates the solvent and cures the binder. This process can be repeated several times to obtain a sufficiently thick and dense carbon layer with good adhesion to the underlying MnO_2 or conductive polymer layers. In some cases, a silicon-based moisture barrier is applied between the carbon layers. The silver layer is then formed on top of the carbon layer by dipping the capacitor elements into an organic solvent-based dispersion of silver particles and organic binder following by air-drying and heat treatment like those with the carbon layer. Figure 2.16 shows tantalum capacitor elements with external carbon and silver layers.

The carbon and silver layers on Fig. 2.16 do not cover the top of the anode to prevent carbon and silver from reaching the dielectric on the tantalum lead wire. As an example of the external layers in solid tantalum capacitor, Fig. 2.17 shows cross section of the tantalum capacitor element with formed tantalum anode (Ta/Ta_2O_5), MnO_2 cathode, and layers of carbon and silver.

External layers of tantalum capacitors, cathode, carbon, and silver, as well as interfaces between these layers, are the major contributors to the ESR of solid tantalum capacitors. In Solid Electrolytic Tantalum capacitors, resistance of the MnO_2 cathode dominates ESR. Some ESR reduction can be achieved by improving the conditions of the manganese nitrite conversion into manganese dioxide; however, the major way to reduce ESR in these capacitors is increasing external surface area of the anodes either by using fluted anodes or multi-anodes [82]. In Polymer Tantalum capacitors with higher conductivity of the PEDOT cathode in comparison to the MnO_2 cathode, the strategy to reduce ESR includes all the external layers and their interfaces as well as the material of the lead frame and adhesive to the lead frame in surface-mount design [83].

Solid tantalum capacitors can be encapsulated either in molded cases for surface-mount pick-and-place assembly on a circuit board or in metal cases for through-hole assembly on a circuit board. In surface-mount design, capacitor elements are connected to the lead frame by welding the anode lead wire to the positive

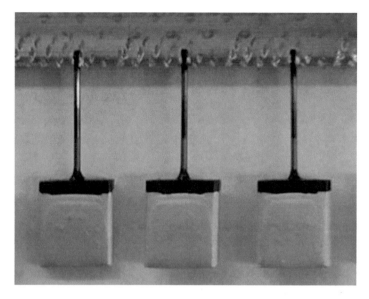

Fig. 2.16 Tantalum capacitor elements with external carbon and silver layers

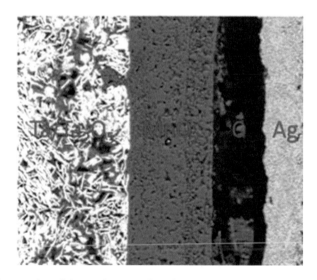

Fig. 2.17 Cross section of the tantalum capacitor element (×400)

termination of the lead frame and attaching the external silver layer to the negative termination of the lead frame with silver adhesive. The carrier bar is then cut away, leaving the capacitor elements attached to the lead frame. The silver adhesive is cured, and the capacitor elements are then molded into an epoxy resin case. As an example, Fig. 2.18 shows general view and cross section of the molded

Fig. 2.18 General view (**a**) and cross section (**b**) of the molded surface-mount tantalum capacitor

surface-mount tantalum capacitor with anode lead wire welded to the positive termination of the lead frame and external silver layer attached to the negative termination of the lead frame with cured silver adhesive.

Some surface-mount tantalum capacitors with MnO_2 cathode have conformal coated plastic encapsulating without a lead frame. These capacitors typically have plated terminations with a short stub of the tantalum lead wire indicating positive termination. Conformal coated design allows for higher volumetric efficiency CV/cm^3; however, external dimensions of this capacitor have more fluctuation in comparison to the molded design, which can be a challenge for the pick-and-place assembly.

Through-hole design is used for Wet Tantalum capacitors and some high-reliability Solid Electrolytic and Polymer Tantalum capacitors. In Wet Tantalum capacitors, formed tantalum anodes are placed in metal cases made with tantalum or other metals and have different internal coatings to increase cathode capacitance. As an example, Fig. 2.19 shows a cross section of a Wet Tantalum capacitor with tantalum case.

In Fig. 2.19 the formed tantalum anode is fixed in the tantalum case with the top and bottom Teflon bushings. The top Teflon bushing is fixed in place by the O-ring pressed in the metal case. Tantalum header with glass insulator is rim welded to the tantalum case. Nickel wire is welded to the tantalum lead wire inside the case and goes through the metallized opening in the center of the glass insulator to provide the capacitor with a positive termination. The cathode tantalum sleeve is sintered to the internal surface of the tantalum case level with tantalum anode. The space between the anode and cathode sleeve is filled with a working electrolyte, which is typically an aqueous solution of sulfuric acid for its high conductivity and low freezing temperature. Nickel wire welded to the bottom of the metal case provides negative termination. Finally, the metal opening in the glass insulator is sealed by solder, and hermeticity is verified by the bubble test and fine leak test.

In hermetic Solid Electrolytic and Polymer Tantalum capacitors, the metal case is typically made of the plated brass, and the capacitor element with the external carbon and silver layers is soldered inside the case. As an example, Fig. 2.20 shows a cross section of the Hermetic Sealed Polymer Tantalum capacitor.

Fig. 2.19 Cross section of the Wet Tantalum capacitor

Fig. 2.20 Cross section of the Hermetic Sealed Polymer Tantalum capacitor

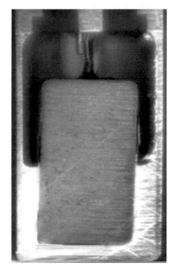

After the encapsulating, tantalum capacitors undergo thermal treatments, aging, and electrical testing, which vary significantly depending on the type of the capacitors. Commercial surface-mount tantalum capacitors undergo several cycles of thermal shock in the temperature range − 55 °C to 125 °C, accelerated aging at 85 °C and voltage exceeding rated voltage, bake out at 125 °C without voltage applied, and IR heat treatment with a temperature profile similar to that of the surface-mount assembly. The electrical parameters such as capacitance, DC leakage, dissipation factor, and ESR are finally tested. When readings of any of these electrical parameters are outside of the specified limits, the capacitors are removed from the population and scrapped.

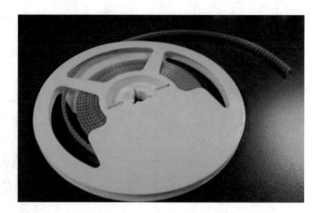

Fig. 2.21 A reel of taped surface-mount tantalum capacitors

The surface-mount and through-hole tantalum capacitors for special applications undergo additional tests such as the Weibull reliability test and the surge test. The qualification of these capacitors also includes a broad range of endurance and environmental tests such as storage and life, ripple current, reverse voltage, thermal stability, and thermal shock, as well as mechanical shock and vibration and solder ability. The conditions and limits of these tests are specified in the manufacturer catalogs for specific types of tantalum capacitors.

The finished and tested surface-mount tantalum capacitors are packed in plastic tape to be shipped to the end users. The plastic tape limits the moisture impact on the capacitors and facilitates pick-and-place assembly on the circuit board. As an example, Fig. 2.21 shows a reel of taped surface-mount tantalum capacitors.

Niobium-based capacitors with either Nb or NbO anodes are manufactured with MnO_2 cathode. The basic technology of these capacitors is similar to the basic technology of the tantalum capacitors with MnO_2 cathode and includes sintering in vacuum of the Nb or NbO anodes, forming and heat treatment of the Nb_2O_5 dielectric, deposition of the MnO_2 cathode inside and outside the porous anodes, applying external carbon and silver layers, encapsulating, and testing. Due to the greater sintering activity of the capacitor-grade niobium powders, sintering temperatures are reduced in comparison to the sintering temperatures with tantalum powders. Additional differences in manufacturing of niobium-based and tantalum capacitors aim to avoid damage of the Nb_2O_5 dielectric, which is more susceptible to degradation in comparison to the Ta_2O_5 dielectric in tantalum capacitors. Particularly, the temperatures of the pyrolytic deposition of the MnO_2 cathode are reduced, and additional reformation steps are added to minimize the field and thermal crystallization and preserve stoichiometric composition of the Nb_2O_5 dielectric. Nevertheless, DCL is increased, and working voltages are reduced in niobium and niobium oxide capacitors in comparison to DCL and working voltages in tantalum capacitors.

2.5   Flawless Dielectric Technology and Simulated Breakdown Screening

Radical improvement in long-term stability and reliability of tantalum capacitors was achieved due to a combination of flawless dielectric technology (F-Tech), providing practically defect-free dielectric to absolute majority of the capacitors, and simulated breakdown screening (SBDS), allowing screening of small percentage of potentially unreliable parts with hidden defects in the dielectric not detectable by traditional DCL testing.

The principle of the F-Tech is based on the fact that density and size of the defects in the Ta_2O_5 dielectric, such as crystalline inclusion, micropores, cracks, etc., are directly proportional to the density and size of the local impurity sites in tantalum anodes prior to anodizing. Mechanical stress during manufacturing, testing, and field application can also cause damage to the dielectric. That is why F-Tech is focused on making tantalum anodes with the highest chemical purity and mechanical robustness.

In tantalum anodes sintered with high-purity capacitor-grade tantalum powders, the impurity sites are typically related to carbide inclusions coming from residuals of organic lubricant mixed with the powder at pressing and oxide inclusions coming from oxygen in native oxide dissolving in the bulk of tantalum particles during their sintering in vacuum. The carbide inclusions in tantalum anodes provoke local thinning and micropores in the dielectric like the ones shown in Fig. 1.5. The oxide inclusions in tantalum anodes provoke crystal growth and cracks in the amorphous matrix of the Ta_2O_5 dielectric like the ones shown in Fig. 1.8.

To prevent contamination of tantalum anodes with carbon, F-Tech uses aqueous de-lubrication instead of traditional thermal de-lubrication by heating pressed anodes in vacuum to decompose and evaporate organic lubricant. During thermal de-lubrication, some of the carbon from the decomposed organic lubricant is absorbed by tantalum particles. Consequently, carbon reacts with tantalum at sintering forming tantalum carbides, which provoke defects in the Ta_2O_5 dielectric during formation of the sintered anodes. In contrast to that, aqueous de-lubrication is performed at room temperature by washing of soluble in water organic lubricant from the pressed tantalum anodes [84]. Aqueous de-lubrication provides low carbon content in the tantalum anodes equal to that in the original powder, while the carbon content increases sharply at the thermal de-lubrication and remains high after the anode sintering (Fig. 2.22).

To minimize oxygen content in tantalum anodes, F-Tech uses deoxidizing by magnesium vapor after the anode sintering in vacuum. Magnesium atoms absorb on the surface of the powder particles and react with oxygen in tantalum, forming magnesium oxide, which is consequently leached from tantalum surface in diluted aqueous solution of sulfuric acid and hydrogen peroxide. Deoxidizing and leaching do not affect mechanical properties of F-Tech anodes typically used in higher-voltage tantalum capacitors and sintered with relatively coarse tantalum powders. As a result of the deoxidizing and leaching, oxygen content in sintered tantalum anodes reduces below oxygen content in the original tantalum powder. As an example, Fig. 2.23 shows typical oxygen content in 50,000 μC/g tantalum powder, sintered with this powder anodes, and after deoxidizing and leaching of the sintered anodes.

F-Tech also provides strong bond between the sintered tantalum powder and tantalum lead wire. Traditionally the lead wire is embedded into the powder during the anode pressing, and then the powder particles are sintered to the wire during the anode sintering in vacuum. With this technology, a mechanical stress during the capacitor manufacturing, testing, and field application can provoke damage of the Ta_2O_5 dielectric on the lead wire, especially, at the anode egress. In F-Tech technology, tantalum anodes are pressed and sintered without the lead wire, and then tantalum wire is welded to the sintered tantalum anode in inert atmosphere. The large welding nugget prevents damage of the Ta_2O_5 dielectric on the lead wire even under the strongest mechanical impact. Figure 2.24 shows cross sections of the tantalum anodes with embedded and welded lead wire.

To purify the welding nugget and surrounding area in the tantalum anodes, final sintering in vacuum is performed after the welding of the tantalum lead wire. The final sintering results in some increase in oxygen content in tantalum anodes due to dissolving of oxygen from the native oxide in the bulk of the tantalum particles; however, final oxygen content in F-Tech anodes remains much lower than oxygen content after the initial sintering due to the preceding steps of the deoxidizing and leaching.

The important part of the F-Tech is the passivation process when tantalum anodes are exposed to air after the final sintering. Uncontrolled passivation and related heat released by oxidation of tantalum particles provoke the thermal oxide growth in the coarser tantalum anodes. As it was shown in Figs. 1.11 and 1.13a, the thermal oxide of tantalum has uneven thickness and high density of crystalline inclusions in amorphous matrix. The thermal oxide partially incorporates into the anodic oxide of

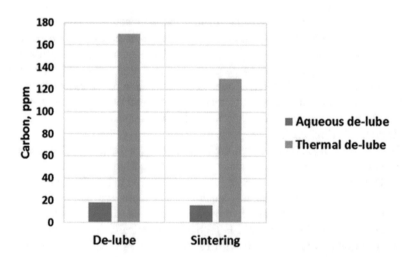

Fig. 2.22 Carbon content in 50,000 μC/g tantalum anodes after de-lubrication and sintering

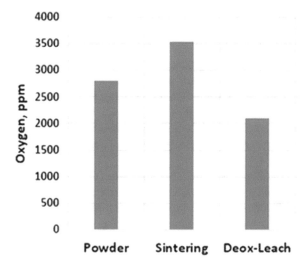

Fig. 2.23 Oxygen content in 50,000 μC/g tantalum powder, sintered with this powder anodes, and after deoxidizing and leaching of the sintered anodes

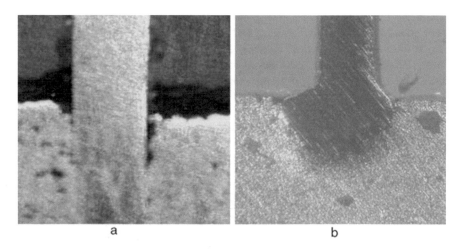

Fig. 2.24 Cross sections of tantalum anodes with embedded (**a**) and welded (**b**) lead wire

tantalum during anodizing, provoking field crystallization and damage of the Ta_2O_5 dielectric. In F-Tech passivation, airflow into the vacuum chamber is controlled to allow heat dissipation and maintain anode temperature close to room temperature. As a result of the controlled passivation, only thin native oxide with amorphous structure grows on tantalum surface that doesn't affect quality of the Ta_2O_5 dielectric in tantalum capacitors.

The F-Tech provides the highest chemical purity and mechanical robustness to the tantalum anodes and, thereby, allows formation of practically flawless Ta_2O_5

dielectric on these anodes. The SEM analysis of a sample of the formed tantalum anodes from every production batch is utilized to make sure that there are no cracks, pores, or any other defects in the Ta_2O_5 dielectric. Figure 2.25 shows the typical SEM image of the flawless Ta_2O_5 dielectric formed on F-Tech tantalum anode.

Tantalum capacitor manufactured with F-Tech anodes demonstrates high DCL stability during long-term testing without de-rating or even at voltages exceeding rated voltage. As an example, Fig. 2.26 shows DCL distribution in D-case 16 µF–25 V Solid Electrolytic Tantalum capacitors manufactured with F-Tech and conventional technology before and after 2000 h of accelerated life test at 1.32 rated voltage and 85 °C.

As one can see in Fig. 2.26a, DCL before the accelerated life test was practically identical in tantalum capacitors with F-Tech and conventional technologies. At the same time after the long-term accelerated life test (Fig. 2.26b), DCL in tantalum capacitors with conventional technology increased on about an order of magnitude, while DCL in tantalum capacitors with F-Tech remained practically unchanged in comparison to DCL before the test.

With any manufacturing technology including F-Tech, there is a probability that a small percentage of the finished capacitors may have hidden defects in their dielectrics, which were not detected during the end-of-line DCL testing and can propagate and cause failures during the field application. Moreover, some harsh accelerated tests like Weibull grading test can induce hidden defects into the dielectric without being detected by existing techniques based on DCL measurements. High DCL at rated voltage indicates flaws in the dielectric, and the capacitors with DCL exceeding the catalog limit are screened at the end-of-line testing; however, low DCL at rated voltage doesn't guarantee flawless dielectric. As an example, Fig. 2.27 shows $I(V)$ characteristics of the two D-case 16 µF–25 V Solid Electrolytic Tantalum capacitors from the same production batch.

Fig. 2.25 SEM image of flawless Ta_2O_5 dielectric formed on F-Tech tantalum anode

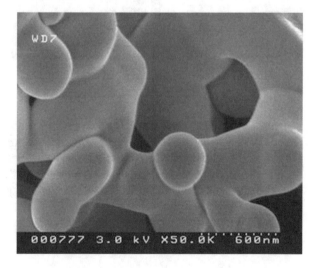

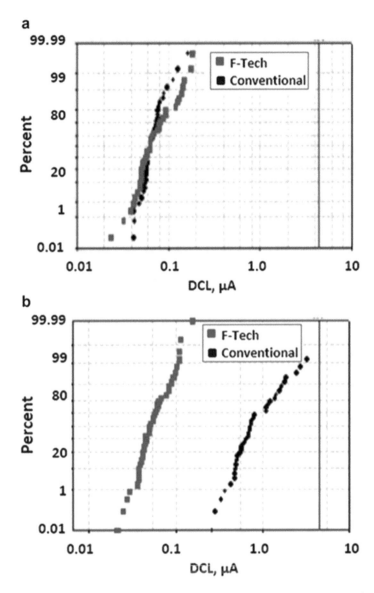

Fig. 2.26 DCL distribution in D-case 16 μF–25 V Solid Electrolytic Tantalum capacitors with F-Tech and conventional technology before (**a**) and after (**b**) 2000 h of accelerated life test

As one can see in Fig. 2.27, these two capacitors have identical DCL at rated voltage (RV) and this DCL is significantly lower than 4 μA DCL limit for this type of tantalum capacitors. When voltage increased above the rated voltage, current in the capacitor A remained low until applied voltage approached formation voltage (FV), which was equal to 88 V. This type of the DCL behavior is typical for the capacitors with high-quality dielectric. In contrast, the current in the capacitor B started increasing rapidly at voltages much lower than formation voltage, which indicates defects in the dielectric. It is obvious that testing the DC leakage current at voltages approaching formation voltage can easily distinguish between the capacitor A and capacitor B; however, this overstress current testing will cause damage to the dielectric in the entire population of the capacitors.

The most efficient way to detect defects in the dielectric of tantalum capacitors is breakdown voltage (BDV) test. Low BDV indicates defects in the dielectric, while high BDV close to formation voltage indicates defect-free dielectric. Despite its efficiency, BDV test is destructive and can't be used for screening purpose. That's why nondestructive simulated breakdown screening (SBDS) was developed, which allows screening of potentially unreliable parts with low BDV without any damage to the population of the capacitors [85, 86]. This test is based on the analysis of the charging characteristic $V(t)$ on tested capacitor C when voltage exceeding average BDV is applied to the tested capacitor and high value series resistor R_s limiting current in the circuit. Figure 2.28 shows principal scheme of the electrical circuit used in SBDS.

The parameters of the screening, $BDV_{average}$ and R_s, are predetermined by measuring the $I(V)$ characteristics on a representative sample from every production batch. The BDV is indicated by the step-like current increase and voltage drop on the capacitor. The series resistance, typically in the range $0.1 \ M\Omega \geq R_s \leq 100 \ M\Omega$, is calculated as average dynamic resistance of the capacitors in vicinity of the average BDV. As an example, Fig. 2.29 demonstrates charge characteristics $V(t)$ of the

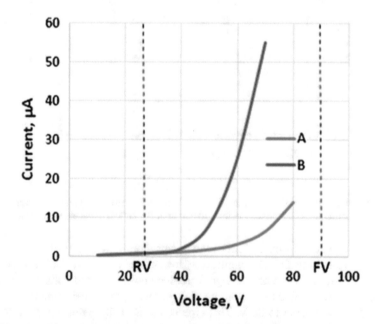

Fig. 2.27 $I(V)$ characteristics of the D-case 16 µF–25 V Solid Electrolytic Tantalum capacitors

capacitors A and B with $I(V)$ characteristics shown in Fig. 2.27. In this case, predetermined parameters were $BDV_{average} = 78$ V and $R_s = 5.2$ MΩ.

According to Fig. 2.29, voltage on the capacitor A reached predetermined BDV average and then was disconnected. In this case, it was only a small voltage drop on the series resistor R_s due to the low current and, thereby, high dynamic resistance of the capacitor A in vicinity of the average BDV. In contrast to that, the voltage on the capacitor B leveled below average BDV and was disconnected after 60 s—maximum duration of the screening. In this case, it was significant voltage drop on the series resistor R_s due to the high current and, thereby, low dynamic resistance of the capacitor B in vicinity of the average BDV. These results show that the final charging voltage at predetermined conditions depends on the DC current in vicinity of the average BDV and can be used for screening purpose as an indicator of the actual BDV in the tested capacitor. The final charging voltages are included in the SBDS distribution built for every production batch. As an example, Fig. 2.30 shows typical SBDS distribution in a production batch of the D-case 16 µF–25 V Solid Electrolytic Tantalum capacitors manufactured with F-Tech.

As one can see in Fig. 2.30, there is normal part of the SBDS distribution with high screening voltages and about 1% "tail" with low screening voltages. The lowest screening voltages in the "tail" typically represent the parts that underwent scintillation breakdown in vicinity of the average BDV when the voltage on the capacitor dropped to zero and didn't increase again before the applied voltage was disconnected at the end of the screening.

High resistance value of the series resistor limiting current through the capacitor and short duration of the high-voltage application prevent any damage to the population of the capacitors during the SBDS, including the parts in the "tail" of the SBDS distribution. Moreover, some self-healing effect takes place so that DCL values at rated voltage after the SBDS become slightly lower in comparison to the DCL values before the SBDS. These DCL values are much lower than DCL limit in this type of the Solid Electrolytic Tantalum capacitors (Fig. 2.31).

Fig. 2.28 Principal scheme of the electrical circuit used in simulated breakdown screening (SBDS)

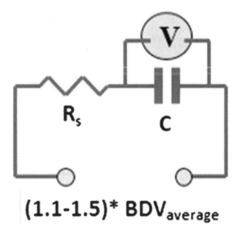

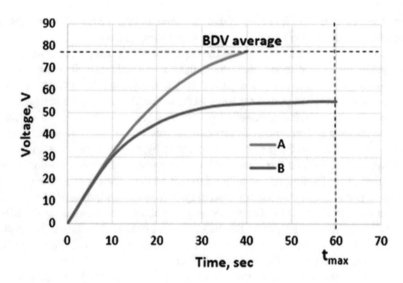

Fig. 2.29 Charge characteristics of the capacitors A and B with $I(V)$ characteristics shown in Fig. 3.28

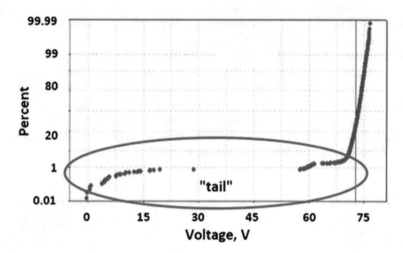

Fig. 2.30 SBDS distribution in D-case 16 µF–25 V MnO_2 tantalum capacitors manufactured with F-Tech

Despite the similarities in DCL between the capacitors from the normal and "tail" parts of the SBDS distribution, these capacitors demonstrate different behaviors at the accelerated life testing. The 2000-h life test of the D-case 4.7 µF–50 V Solid Electrolytic Tantalum capacitors at 85 °C and 1.32 rated voltage didn't cause any failures or noticeable increase in DCL in the capacitors from the normal part of the SBDS distribution, while about 30% of the capacitors from the "tail" failed short and blackened during the test. Figure 2.32 shows circuit boards with the capacitors from the normal and "tail" parts of the SBDS distribution after the accelerated life test.

As one can see in Fig. 2.32, there are no damaged capacitors on the circuit board with the capacitors from the normal part of the SBDS distribution (Fig. 2.32a), while a substantial number of the capacitors from the "tail" have signs of damage (Fig. 2.32b). These results confirm that the "tail" in the SBDS distribution identifies unreliable capacitors, which have normal DCL at rated voltage and, thereby, can't be screened by traditional DCL techniques. The SBDS allows to identify and remove a small percentage of the unreliable capacitors with hidden defects in the dielectric from the normal population of the tantalum capacitors.

To verify efficiency and nondestructive nature of the SBDS, BDV test was performed on the samples of the D-case 4.7 μF–50 V Solid Electrolytic Tantalum capacitors from the same F-Tech production batch before SBDS and after SBDS and removal of the "tail" (Fig. 2.33).

According to Fig. 2.33, a small percentage of the capacitors with low BDV were removed as the "tail" in the SBDS distribution, while no damage (no change in BDV) was done to the entire population of the capacitors as a result of the SBDS. The long-term experience in manufacturing, testing, and field application of tantalum capacitors showed that for reliable performance, BDV in these capacitors should be at least two times higher than working voltage. As one can see in Fig. 2.33, all the capacitors with rated voltage 50 V, F-Tech, and SBDS have BDV exceeding 100 V and can be used without de-rating, while before SBDS a few capacitors had BDV below 100 V and needed de-rating for reliable performance.

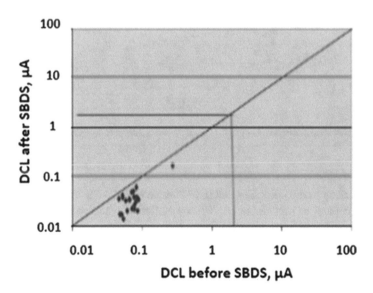

Fig. 2.31 DCL in D-case 16 μF–25 V Solid Electrolytic Tantalum capacitors before and after SBDS

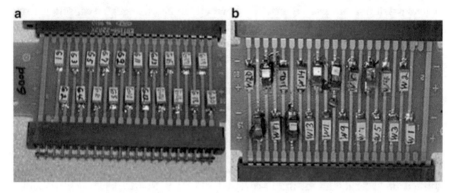

Fig. 2.32 Circuit board with D-case 4.7 μF–50 V Solid Electrolytic Tantalum capacitors from normal (**a**) and "tail" (**b**) parts of the SBDS distribution after the accelerated life test

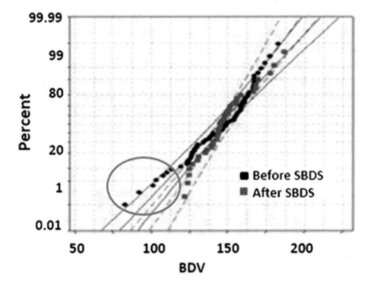

Fig. 2.33 BDV distributions in D-case 4.7 μF–50 V Solid Electrolytic Tantalum capacitors before and after SBDS

Comparison of the reliability and failure mode in Solid Electrolytic and Polymer Tantalum capacitors manufactured with conventional technology and F-Tech/SBDS will be discussed in Sect. 3.2 of this book.

Chapter 3
Applications

The results presented in this chapter are dedicated to the specific charge and energy of tantalum capacitors as a function of formation voltage with emphasis on losses at high and low formation voltages; de-rating and high-reliability approach for all types of tantalum capacitors; high working voltage, asymmetric conduction, and anomalous currents with emphasis on Polymer Tantalum capacitors; and high-temperature applications of MnO_2 tantalum capacitors.

3.1 High Charge and Energy Efficiency

The major way to increase charge volumetric efficiency (CV/cm^3) and weight efficiency (CV/g) in tantalum capacitors is by using finer tantalum powders with smaller primary particles, thereby increasing the specific surface area of tantalum anodes. Originally produced coarsest tantalum powders obtained by electron-beam purification and crushing of tantalum ingot (EB powders) have primary particles with an average size in the tens of microns and a CV/g equal or below 10,000 μC/g. The finest currently produced tantalum powders obtained by sodium reduction of potassium tantalum fluoride or direct magnesium reduction of tantalum pentoxide have submicron primary particles and CV/g equal or above 250,000 μC/g. Figure 3.1 shows the evolution of tantalum powders and images of tantalum capacitors with the same capacitance ($C = 330$ μF) and working voltage ($W_V = 6.3$ V) made with coarse, medium, and high CV tantalum powders [87]. The size of the dots in Fig. 3.1 is proportional to the average size of the primary particles in tantalum powder.

Usage of the high CV tantalum powders in tantalum capacitors is limited to low-voltage capacitors with thin oxide dielectrics. When the dielectric thickness increases due to the higher formation voltages, CV/g of tantalum anodes with high

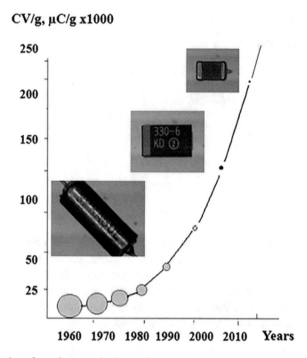

CV/g, µC/g x1000

Fig. 3.1 Evolution of tantalum powder in tantalum capacitors

CV powder rapidly decreases, while *CV*/g of tantalum anodes with coarser powder remains constant in a broad range of formation voltages. As an example, Fig. 3.2 shows *CV*/g dependence on formation voltage in tantalum anodes with 150,000 and 50,000 µC/g tantalum powder.

For parallel plate capacitors, the capacitance *C* is given as:

$$C = kk_o A / t,$$

where *k* is the dielectric constant, k_o the permittivity of vacuum, *A* the surface area, and *t* the dielectric thickness. The dielectric thickness is directly proportional to the formation voltage V_f:

$$t = aV_f,$$

where the proportionality coefficient *a* is termed voltage constant [17]. Assuming that the dielectric constant does not change with formation voltage, *CV*/g at given temperature should remain constant as a function of formation voltage; however, presented in Fig. 3.2 experimental results show *CV*/g to decrease at both high and low formation voltages.

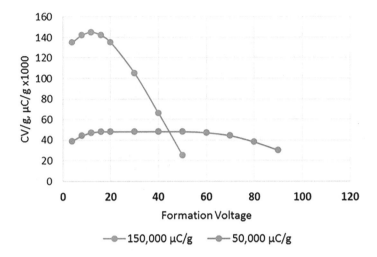

Fig. 3.2 *CV*/g dependence on formation voltage in tantalum anodes with 150,000 and 50,000 μC/g tantalum powder

The high-voltage *CV* loss in tantalum anodes was explained by the neck-and-pore model [88, 89]. During formation, the anodic oxide film of tantalum is partially growing into the initial tantalum surface, consuming tantalum, and partially outward from the initial tantalum surface due to the lower density of tantalum oxide compared to that of tantalum. The ratio between the inward and outward growth of the anodic oxide film is about 2:3 on a flat tantalum substrate and large tantalum particles, increasing as the radius of tantalum particles gets smaller. The inward growth of the anodic oxide film of tantalum reduces the radius of the sintered tantalum particles and eventually disrupts the current flow in the thinnest parts of the porous tantalum anodes—the necks connecting the powder particles. The outward growth of the anodic oxide film of tantalum eventually closes the pores between the powder particles so that the cathode cannot get inside during further steps of the capacitor manufacturing. Both processes, consumption of tantalum and closure of pores, decrease surface area and, thereby, *CV* of the tantalum anodes. As an example, Fig. 3.3 shows breakage of the tantalum anode sintered with 50,000 μC/g tantalum powder and formed to 75 V [87].

As one can see in Fig. 3.3, some remaining necks between the powder particles (light areas) are very small or even totally consumed by the anodic oxide film of tantalum (gray areas) and some pores between the particles are very small or totally clogged by the oxide film. There is already *CV* loss in this anode at 75 V formation, which becomes greater at higher formation voltages (Fig. 3.2). The size of the necks connecting the powder particles can be increased by higher press density and sintering temperature of the tantalum anodes; however, this will result in higher shrinkage (Fig. 2.7) and, thereby, smaller pores between the powder particles. At lower press

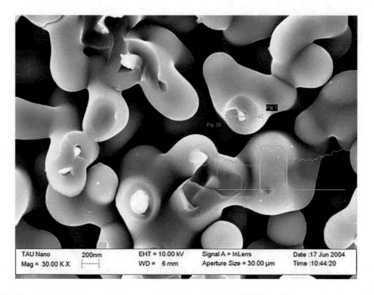

Fig. 3.3 Breakage of the tantalum anode sintered with 50,000 μC/g tantalum powder and formed to 75 V

density and sintering temperature, the pores will be more open, but the necks connecting the powder particles will be too small, resulting in rapid *CV* loss with formation voltage and poor mechanical strength of the anodes. The common practice in manufacturing of tantalum anodes is to optimize press density and sintering temperature to achieve maximum *CV*/g and acceptable mechanical strength for given type of the tantalum powder.

The bond between the powder particles and the lead wire is also a limiting factor in utilization of the high *CV* tantalum powder. Sintering activity between the powder particles with small radius and the wire with much larger radius is reduced in comparison to the sintering activity between the fine powder particles [65]. To achieve sufficient powder-wire bond, the sintering temperature should be increased, which results in higher anode shrinkage and surface area loss. To address this issue, the press density can be increased locally in small area around the lead wire. In this case, the powder-wire bond is improved without sintering temperature increase and a significant *CV*/g loss [90].

One more limiting factor in the utilization of the high *CV* tantalum powder in tantalum capacitors is the amount of oxygen in tantalum anodes sintered with these powders [81, 87]. As the powder *CV* increases, the amount of oxygen in this powder also increases though the sintering temperature decreases to avoid over-shrinkage and loss of surface area in the sintered anodes (Fig. 3.4).

The increase in total oxygen content in higher *CV* tantalum powders is due to the larger surface-to-volume ratio in the smaller primary particles and consequential

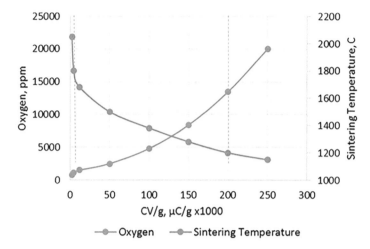

Fig. 3.4 Oxygen content and sintering temperature as a function of the *CV*/g of tantalum powder

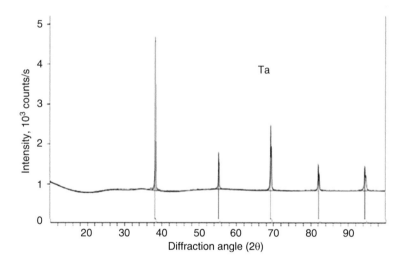

Fig. 3.5 X-ray diffraction pattern of the 200,000 µC/g tantalum powder

larger effect of oxygen in the native oxide on the particle surface. At the same time, bulk oxygen content in the higher *CV* tantalum powders remains low due to the deoxidizing included in the powder manufacturing. The bulk oxygen content in tantalum powder can be identified with X-ray diffraction analysis (XRD) through the period of crystalline lattice of tantalum, which increases with larger amounts of oxygen dissolved in tantalum as an interstitial impurity [48]. As an example, Fig. 3.5 shows XRD pattern of tantalum powder with specific charge 200,000 µC/g and oxygen content about 15,000 ppm.

According to Fig. 3.5, the powder consists of tantalum phase with cubic crystalline structure and lattice parameter $a = 3.3088$ Å. This lattice parameter is close to the lattice parameter in pure tantalum [48], indicating that the bulk of the powder particles is free of oxygen and that practically all the oxygen in this powder is in the native oxide on the surface of the powder particles.

The anodes sintered with tantalum powder with specific charge 200,000 μC/g have oxygen content about 20,500 ppm and diffraction pattern significantly different from diffraction pattern of the tantalum powder before sintering (Fig. 3.6).

According to Fig. 3.6, the sintered anode consists of two phases: tantalum phase with lattice parameter $a = 3.3192$ Å, indicating a saturated solid solution of oxygen in tantalum, and tantalum pentoxide phase. The inclusions of the crystalline tantalum oxide were also detected by scanning electron microscopy (SEM) on the surface and inside the tantalum anodes sintered with 200,000 μC/g tantalum powder (Fig. 3.7).

Similarly, as shown earlier in Fig. 1.20, the harder oxide particles in the much-softer sintered tantalum anode affect its mechanical properties, provoking anode cracking and chipping during manufacturing, testing, and in the field when mechanical stress is applied to the anodes. Post-sintering deoxidizing of the high CV tantalum anodes performed similarly to the process used in the manufacturing of the tantalum powder will reduce oxygen content in these anodes [91]; however, it will also reduce crush strength of the anodes and pull strength of the lead wire embedded in the anodes. As an example, Fig. 3.8 shows a disconnection between the powder and the embedded lead wire in a breakage of tantalum anode sintered in vacuum with 200,000 μC/g tantalum powder and then deoxidized by magnesium vapor.

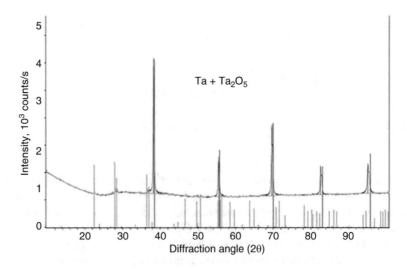

Fig. 3.6 X-ray diffraction pattern of tantalum anode sintered with 200,000 μC/g tantalum powder

Fig. 3.7 SEM image of
the crystalline tantalum
pentoxide in tantalum
anode sintered with
200,000 μC/g tantalum

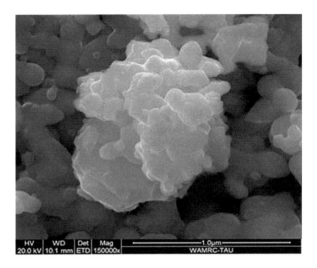

Fig. 3.8 A breakage of
tantalum anode sintered
with 200,000 μC/g
tantalum powder and then
deoxidized by magnesium
vapor

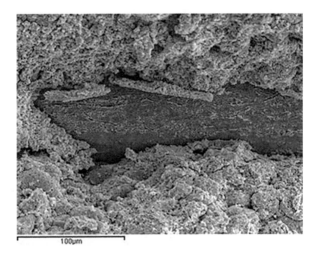

The loss in the mechanical strength of the high CV tantalum anodes as a result of deoxidizing by magnesium vapor evidences that a significant part of the bonds between the powder particles and between the particles and the lead wire in these anodes is made of tantalum oxide. Deoxidizing transforms tantalum oxide into porous tantalum with higher density and, thereby, lowers volume in comparison to the volume and density of the tantalum oxide. This transformation destroys mechanical bonds between the powder particles and between the particles and wire, making impractical deoxidizing of the high CV tantalum anodes after their sintering in vacuum.

Mechanically strong tantalum anodes with record high charge and energy efficiency can be obtained by sintering of the high *CV* tantalum powder combined with
deoxidizing by magnesium vapor (deox-sintering) [66, 87, 90, 92–96]. Morphology
of the deox-sintered anodes is very different from morphology of anodes sintered in
vacuum with the same tantalum powder. As an example, Fig. 3.9 shows SEM images
of the tantalum anodes with 200,000 µC/g tantalum powder sintered in vacuum,
deox-sintered, and deox-sintered with double the amount of magnesium and time in
comparison to these in regular deox-sintering.

As one can see in Fig. 3.9, deox-sintered tantalum anodes have larger size particles in comparison to the size of the particles in the original powder, and these particles become larger with increasing the amount of magnesium and time. The
deox-sintered anodes have thick uniform necks connecting the powder particles and
large pores between these particles. The combination of the thick necks and large
pores cannot be achieved by sintering of the tantalum powder in vacuum due to

Fig. 3.9 SEM images of tantalum anodes with 200,000 µC/g powder sintered in vacuum (**a**),
deox-sintered (**b**), deox-sintered with double the amount of magnesium and time (**c**)

over-shrinkage and reduction in the pore size at the high sintering temperatures required to grow thicker necks between the powder particles. Contrary to the sintering in vacuum, deox-sintered anodes have very little or no shrinkage. The oxygen content in the deox-sintered anodes is lower than the oxygen content in the tantalum powder they were sintered with, while the oxygen content in anodes sintered in vacuum with high CV/g tantalum powder is much higher than the oxygen content in the original powder.

The differences in morphology, oxygen content, and shrinkage between the sintered in vacuum and deox-sintered tantalum anodes are related to the differences in the dominant sintering mechanisms in these processes. During sintering in vacuum, the dominant sintering mechanism is the bulk diffusion of tantalum which causes mutual penetration of the powder particles and shrinkage of the anode. Deox-sintering is performed at temperatures much lower than temperatures used for sintering in vacuum of the same type of the tantalum powder. At deox-sintering temperatures, the bulk diffusion of tantalum, which is exponentially dependent on temperature, is practically negligible and the dominant sintering mechanism becomes the surface diffusion of tantalum. In this case, tantalum atoms diffuse on the surface of the powder particles to the points with the smallest radius—the necks between adjacent particles—building up the necks until their radius becomes comparable to the radius of the powder particles. In this process, the finer fraction of the primary particles becomes a building material for the necks between the larger primary particles, forming thick uniform necks between these particles and leaving large open pores inside the porous anodes. This mechanism of tantalum redistribution during the deox-sintering process does not require shrinkage to build a mechanically strong anode as it takes place during sintering in vacuum of the tantalum powder.

Since oxygen in tantalum is an inhibitor to sintering, surface diffusion of tantalum as a sintering mechanism is efficient only when oxygen is removed from the tantalum powder during the sintering process. The other condition for the surface diffusion of tantalum as the dominant sintering mechanism is the presence of the low-radius initial necks between the powder particles that work as drains for the diffusion of tantalum atoms. Deox-sintering satisfies both conditions since it is performed in a deoxidizing atmosphere of magnesium vapor and the powder is pressed into pellets prior to the deox-sintering.

Due to the differences in morphology between the sintered in vacuum and deox-sintered tantalum anodes, the specific charge dependence on formation voltage is also different in these anodes. As an example, Fig. 3.10 shows CV/g vs. formation voltage in tantalum anodes with 200,000 µCV/g tantalum powder sintered in vacuum, deox-sintered, and deox-sintered with double the amount of magnesium and time (the morphology of these anodes is shown in Fig. 3.9).

As one can see from Fig. 3.10, at low formation voltages, CV/g in deox-sintered tantalum anodes is lower than CV/g in sintered vacuum tantalum anodes, which correlates with, as shown in Fig. 3.9, larger particle size and, thereby, lower surface area in deox-sintered anodes compared to sintered in vacuum tantalum anodes. At the same time, at higher formation voltages, CV/g in deox-sintered tantalum anodes becomes higher than CV/g in sintered vacuum tantalum anodes. The advantage in

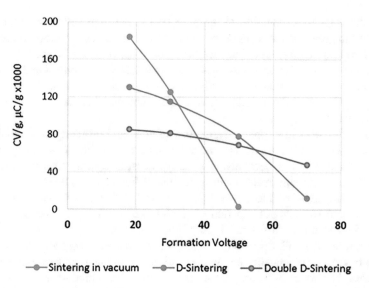

Fig. 3.10 *CV*/g vs. formation voltage in tantalum anodes with 200,000 μC/g powder sintered in vacuum (a), deox-sintered, (b) deox-sintered with double the amount of magnesium and time (c)

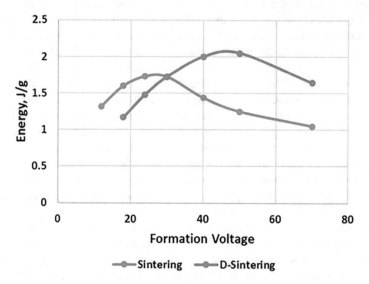

Fig. 3.11 *E/g* vs. formation voltage in tantalum anodes sintered in vacuum with traditionally used tantalum powders and tantalum anodes deox-sintered with 200,000 μCV/g tantalum powder

CV/g in the deox-sintered tantalum anodes is due to these anodes' morphology that combines thick uniform necks between the powder particles and with large open pores inside the anodes. This unique morphology allows deox-sintered anodes to achieve not only the highest specific charge in the intermediate range of formation voltages but also the highest specific energy $E = CV^2/2$ in comparison to tantalum anodes sintered in vacuum with any tantalum powder (Fig. 3.11).

According to Fig. 3.11, specific energy of both sintered in vacuum and deox-sintered tantalum anodes increases with formation voltage at low formation voltages, reaches a maximum in the intermediate range of formation voltages, and then decreases at higher formation voltages due to rapid CV/g loss. Maximum specific energy of the deox-sintered tantalum anodes is shifted to higher formation voltages and exceeds the maximum specific energy of the sintered in vacuum tantalum anodes. Due to the high chemical purity, the formed deox-sintered anodes provide practically defect-free dielectric like the one shown in Fig. 2.25. The flawless dielectric technology using deox-sintered anodes (F-Tech with deox-sintering) also includes high density powder region around the embedded in the anode wire and distortion of the wire at the anode pressing to improve the powder-wire bond in the deox-sintered anodes [97].

One of the trends in modern electronics is reducing the operating voltages to minimize the heat dissipation and increase component density on the circuit board. Besides CV/g roll-off with increasing formation voltage, which is more pronounced with higher CV/g tantalum powders, there is also CV/g roll-off with decreasing formation voltage below approximately 12 V. This low-voltage CV/g loss limits the efficiency of tantalum capacitors at lower-voltage applications and is often accompanied by increases in the DC leakage (DCL). As an example, Fig. 3.12 shows the CV/g dependence on formation voltage in tantalum anodes with 40,000 and 150,000 $\mu C/g$ tantalum powders and the CV/g loss, calculated as the percent difference relative to the CV/g at formation voltage 12 V [49].

According to Fig. 3.12, CV/g decreases with decreasing formation voltage for both powders. The absolute value of the low-voltage CV/g loss is higher with the

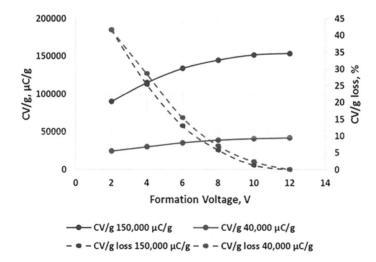

Fig. 3.12 CV/g and CV/g loss vs. formation voltage in tantalum anodes with 40,000 and 150,000 $\mu C/g$ tantalum powders

150,000 µC/g tantalum powder in comparison to the 40,000 µC/g tantalum powder. At the same time, the relative CV/g loss is practically identical for these two tantalum powders.

The electrical measurements in [49] were combined with the transmission electron microscopy (TEM) analysis of thin sections of tantalum anodes transparent for electron beam. For anodic oxide thickness measurements, multiple photomicrographs were taken at the same magnification from different areas for each sample. Figure 3.13 presents the thickness of oxide film on tantalum anodes with 40,000 µCV/g tantalum powder as a function of formation voltage.

As shown in Fig. 3.13, the thickness of the oxide film increases linearly with increasing formation voltage with the slope 2.3 nm/V, which is in good agreement with the slope presented by Young for formation of Ta_2O_5 at 80 °C [17]. The y-intercept, t_o, obtained from Fig. 3.13 occurs at 3.3 nm, which is equal to the thickness of native oxide measured by TEM on tantalum surface prior to anodizing. From Fig. 3.13, TEM thickness t of the oxide film on tantalum anodes as a function of formation voltage V_f is equal to:

$$t = t_o + aV_f,$$

where $t_o = 3.3$ nm and $a = 2.3$ nm/V for formation at 80 °C.

The experimental data show that the thickness of the native oxide is added to the thickness of the anodic oxide, increasing the total thickness of the dielectric on tantalum anodes and, thereby, causing the loss of capacitance and CV/g. This CV/g loss is significant at low formation voltages when the thickness of the native oxide on tantalum is comparable with the thickness of the anodic oxide of tantalum. The addition of native oxide to the thickness of the native oxide of tantalum without a

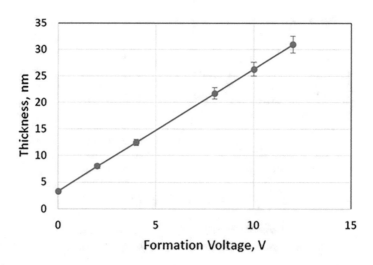

Fig. 3.13 Thickness of oxide film on tantalum anodes with 40,000 µCV/g tantalum powder as a function of formation voltage

noticeable voltage drop may be caused by the tunneling through the very thin dielectric.

Heat treatment in air at 300 °C for 1 h was performed on tantalum anodes sintered in vacuum with 40,000 µCV/g tantalum powder (tantalum anodes sintered with 150,000 µCV/g tantalum powder would ignite at this temperature). The TEM thickness and capacitance for heat-treated and non-heat-treated tantalum anodes are shown in Fig. 3.14.

The experimental results presented in Fig. 3.14 demonstrate that once the initial native oxide thickness is exceeded, the anodic oxide growth is identical for heat-treated and non-heat-treated tantalum anodes regardless of the initial thermal oxide thickness. The thermal oxide gradually incorporates into the growing anodic oxide and eventually doesn't affect the total oxide thickness and capacitance of the formed tantalum anodes.

Heat treatment in air at 300 °C for 1 h was also performed after formation of tantalum anodes sintered in vacuum with 40,000 µCV/g and 150,000 µC/g tantalum powders. Figure 3.15 shows the effect of post-formation heat treatment on CV/g of these anodes.

As shown in Fig. 3.15, post-formation heat treatment in air results in additional CV/g loss, particularly at low formation voltages. This additional loss is much greater in the tantalum anodes sintered in vacuum with the 150,000 µC/g powder than those with the 40,000 µC/g powder. TEM analysis showed a change in oxide thickness in formed tantalum anodes due to heat treatment in air, which correlated with capacitance loss and thereby CV/g loss shown in Fig. 3.15. In this case, the increase in oxide thickness can be attributed to oxygen diffusion from air through the anodic Ta_2O_5 film toward the oxide-tantalum interface, causing the growth of an additional thermal oxide on this interface. The thermal oxidation rate will decrease with increasing thickness of the anodic oxide (increasing formation voltage) because the diffusion length for oxygen becomes larger. Thermal oxidation is also diminished in tantalum anodes with coarser tantalum powder having lower bulk oxygen

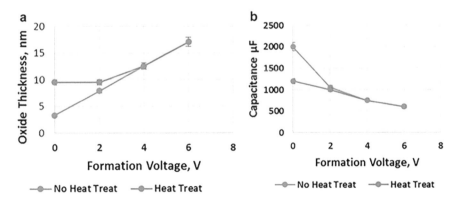

Fig. 3.14 Effect of preformation heat treatment in air on oxide thickness (**a**) and capacitance (**b**) of tantalum anodes sintered in vacuum with 40,000 µC/g tantalum powder

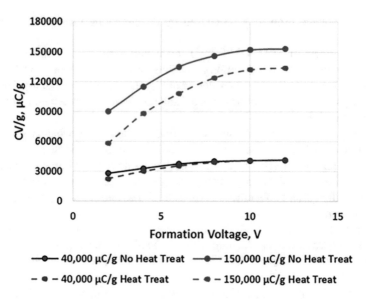

Fig. 3.15 Effect of post-formation heat treatment in air on *CV*/g of tantalum anodes sintered in vacuum with 40,000 and 150,000 μC/g tantalum powders

content. In this case, oxygen atoms reaching the oxide-tantalum interface dissolve in the undersaturated tantalum volume instead of forming a thermal oxide.

In conclusion, usage of high *CV*/g tantalum powders is limited to low-voltage tantalum capacitors with low formation voltages. The *CV*/g loss at higher formation voltages is caused by reducing the surface area of the anode due to the reduction of the radius of the powder particles, consumption of the necks connecting powder particles, and the closure of pores between particles by the anodic oxide film of tantalum growing inward and outward from the original tantalum surface. The *CV*/g loss with decreasing formation voltage is a consequence of the addition of native oxide thickness to the thickness of the anodic oxide of tantalum, which is more pronounced at low formation voltages due to the greater contribution of native oxide thickness to the total dielectric thickness.

Heat treatment of the tantalum anodes in air before formation results in additional thermal oxide growth on the tantalum surface. During formation, the thermal oxide thickness incorporates into the growing anodic oxide film except for the thickness of the initial native oxide of tantalum. Though the total oxide thickness and capacitance of the heat-treated in air tantalum anodes are not affected by the thermal oxide growth, its incorporation into the anodic oxide can cause DCL increase and DCL instability with time due to the large concentration of the crystalline inclusions in the bulk of the thermal oxide of tantalum. That is why it is critically important for the capacitor quality to prevent temperature increase and the thermal oxide growth as a result of the exothermic oxidation during tantalum anodes exposure to air after their sintering in vacuum. High *CV*/g tantalum anodes are more prone to the thermal oxidation and even ignition due to the high surface-to-volume ratio in the fine powder particles.

Heat treatment of the tantalum anodes in air after low-voltage formation results in additional *CV* loss due to the increase in oxide thickness caused by the diffusion of oxygen from air through the thin oxide film and toward the oxide-Ta interface where growth of additional thermal oxide takes place. The thermal oxide growth is greater in higher *CV*/g tantalum anodes, whose higher bulk oxygen content prevents them from serving as a sink for additional oxygen. This effect limits the application of the highest *CV* tantalum powders in tantalum capacitors with MnO_2 cathodes deposited pyrolytically at temperatures approaching 300 °C. In contrast, Polymer Tantalum capacitors with conductive polymer cathodes do not require high-temperature heat treatment in air and, thereby, can be manufactured with the highest *CV*/g tantalum powders.

The critical factor limiting usage of the high *CV* tantalum powders is oversaturation of tantalum with oxygen and precipitation of the tantalum oxide inclusions in sintered in vacuum tantalum anodes, which affect electrical and mechanical properties of tantalum capacitors manufactured with these anodes. Sintering of the high *CV*/g tantalum powder in a deoxidizing atmosphere (deox-sintering) allows for manufacturing of tantalum anodes with low oxygen content, high mechanical strength, record high charge and energy efficiency and flawless dielectric in the intermediate range of formation voltages. These properties are provided by high chemical purity and a unique morphology of the deox-sintered anodes that combines thick uniform necks connecting the powder particles and large open pores between these particles.

3.2   Reliability and Failure Mode in Solid Tantalum Capacitors

Since the 1970s, a Weibull approach has been used to assess reliability of Solid Electrolytic Tantalum capacitors. The recent updates of this approach are reported in [98]. Accordingly, Solid Electrolytic Tantalum capacitors are tested at temperature 85 °C and applied voltage (V_a) exceeding rated voltage (V_r). The acceleration factor (*A*) for this test is calculated as:

$$A \approx 7.034 \times 10^9 \exp\left(18.77 \frac{V_a}{V_r}\right)$$

The Weibull equation is based on the results of the accelerated testing of numerous Solid Electrolytic Tantalum capacitors from different manufacturers with different values of capacitance and rated voltages. The characteristic feature of Solid Electrolytic Tantalum capacitors is a decreasing cumulative failure rate with time ($\beta < 1$, where β is the slope of the cumulative failure rate versus time).

The time-to-failure distribution at accelerated testing of Polymer Tantalum capacitors is different from that in Solid Electrolytic Tantalum capacitors [99, 100]. Typically, there are no or very few failures at the earlier stages of the testing and then majority of the capacitors fail within relatively short period of time (wear-out region). The acceleration factor in Polymer Tantalum capacitors was calculated in [99, 100] using Prokopovich-Vaskas equation:

$$A = \left(\frac{V_a}{V_r}\right)^n \exp\left[E_a\left(\frac{1}{T_1} - \frac{1}{T_2}\right)/k\right]$$

where T_1 and T_2 are absolute temperatures at the accelerated testing and at application correspondingly, k is Boltzmann constant $k = 1.380649 \times 10^{-23}$ J/K, and exponent n and activation energy E_a determined from the median time to failure (50% failed) dependence on voltage and temperature. Though as presented in [99, 100], median times to failure in Polymer Tantalum capacitors at normal application conditions were lengthy, the wear-out performance of these capacitors was concerning, especially, in high-reliability applications.

The typical failure mode in Solid Electrolytic and Polymer Tantalum capacitors is low insulation resistance or a short. The hypothesis about ignition and burning tantalum failure mode in SMD-type Solid Electrolytic Tantalum capacitors was initially presented in [101] after many years of broad applications of these capacitors without any ignition and burning reported. According to [101] the heat released at breakdown of these capacitors causes microcracks in the Ta_2O_5 dielectric where exposed tantalum anode is ignited by oxygen from the reducing MnO_2 cathode. Since PEDOT cathode doesn't have active oxygen in its molecular structure, no ignition and burning of tantalum anodes takes place at the breakdown event in Polymer Tantalum capacitors (benign failure mode).

To prove the hypothesis about ignition and burning tantalum failure mode in Solid Electrolytic Tantalum capacitors and no ignition failure mode in Polymer Tantalum capacitors, the reverse polarity voltage was continuously increased on these capacitors until Solid Electrolytic Tantalum capacitors ignited at reverse voltage about the twice rated voltage and the voltage was disconnected. It is well known that tantalum capacitors are polar with breakdown voltage at reverse polarity much less than the rated voltage. In this case the ignitions reported in [101] could be caused by flammable epoxy compound in external encapsulation and could be observed on Polymer Tantalum capacitors as well with slight additional increase of the reverse voltage and related current and heat.

To reduce the risk of failure in Solid Electrolytic Tantalum capacitors, 50% de-rating ($V_a/V_r = \frac{1}{2}$) was recommended [102] and accepted for most applications. The 50% de-rating causes about 10× loss in volumetric efficiency due to a combination of the thicker dielectric and lower anode surface area since larger tantalum particles, coarser tantalum powders, are needed to form the thicker dielectric. As an example, Fig. 3.16 shows D-case Solid Electrolytic Tantalum capacitor with capacitance 4.7 μF and rated voltage 50 V, which can be used with maximum working voltage

Fig. 3.16 Solid
Electrolytic Tantalum
capacitors: D-case 4.7 µF,
50 V (*left*), and A-case
4.7 µF, 25 V (*right*)

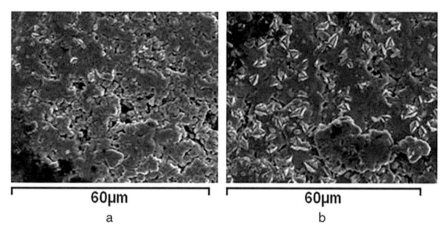

Fig. 3.17 SEM images of tantalum anodes formed at formation voltages 130 V (**a**) and 190 V (**b**)

25 V due to the 50% de-rating, and A-case Solid Electrolytic Tantalum capacitor
with the same capacitance 4.7 µF and rated voltage 25 V (no de-rating). The ratio in
volume between these two capacitors is about 10:1.

Even though the de-rating reduces the failure rate in finished tantalum capacitors, de-rating approach to the capacitor technology may have negative effect on the
performance and reliability of these capacitors, especially, at higher working voltages. The Ta_2O_5 dielectric is thicker in the capacitors with de-rating in comparison
to the thickness of the dielectric in non-de-rated capacitors with the same working
voltage. As it was discussed earlier, thicker Ta_2O_5 dielectrics formed at higher formation voltages are more susceptible to field and thermal crystallization of the
amorphous matrix in comparison to the thinner Ta_2O_5 dielectrics formed at lower
formation voltages. As an example, Fig. 3.17 shows SEM images of the surface of
conventional tantalum anodes formed at formation voltages 130 and 190 V.

As one can see in Fig. 3.17, there are just a few small cracks from the growing crystals in thinner dielectric formed at lower formation voltage, while there are numerous larger cracks in thicker dielectric formed at higher formation voltage. The initial crystals progress during manufacturing, especially, in Solid Electrolytic Tantalum capacitors with pyrolytic deposition of the MnO_2 cathode. Because of the damage to the Ta_2O_5 dielectric due to the crystallization of the amorphous matrix, breakdown voltage in Solid Electrolytic Tantalum capacitors with formation voltage 190 V was lower than breakdown voltage in Solid Electrolytic Tantalum capacitors with formation voltage 130 V, despite the fact the thickness of the Ta_2O_5 dielectric was larger at higher formation voltage (Fig. 3.18).

The loss in volumetric efficiency and fear of ignition and burning tantalum failure mode, which dominated online publications, resulted in decline in general applications of Solid Electrolytic Tantalum capacitors including the applications where high reliability and environmental stability of these capacitors are most needed.

The reliability and failure mode in SMD-type Solid Electrolytic and Polymer Tantalum capacitors was investigated on the parts manufactured with conventional technology and flawless dielectric technology (F-Tech) [103]. As it was presented in Sect. 2.5, F-Tech suppresses typical defects such as crystalline inclusions in the amorphous matrix of the Ta_2O_5 dielectric that continue growth during the capacitor testing and field applications eventually causing cracks in the dielectric and capacitor failures. The critical part of the F-Tech is reducing bulk oxygen content in tantalum anodes that promotes formation of the crystalline seeds at the anode-dielectric interface. Reducing the oxygen content in tantalum anodes in F-Tech can be performed either by treatment of the sintered in vacuum tantalum anodes in deoxidizing atmosphere, such as magnesium vapor, or by sintering of the pressed tantalum powder pellets in a deoxidizing atmosphere (F-Tech with deox-sintering).

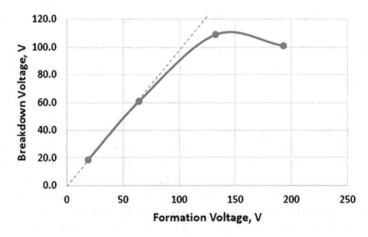

Fig. 3.18 Breakdown voltage dependence on formation voltage in Solid Electrolytic Tantalum capacitors

Accelerated Weibull test of X-case 6.8 μF, 50 V Solid Electrolytic Tantalum capacitors manufactured with conventional technology and F-Tech was performed at 70 V ($V_a/V_r = 1.4$) and 85 °C. The cumulative percent of failed vs. time and failure rate vs. time in these capacitors are shown in Fig. 3.19. According to [98], the failure rate was calculated for 2 and 40 h of testing using experimental values of the cumulative percent of failed at these time intervals and acceleration factor $A = 1824.38$ according to the Weibull equation with $V_a/V_r = 1.4$.

As one can see in Fig. 3.19a, the percent of failed parts and failure rates are lower in capacitors manufactured with F-Tech in comparison to the capacitors manufactured with conventional technology. Significant reduction in the failure rate in Solid

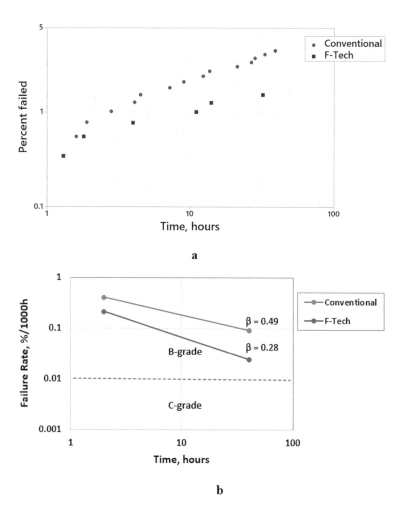

Fig. 3.19 Cumulative percent of failed (**a**) and failure rate (**b**) versus time at Weibull test at 70 V and 85 °C of X-case 6.8 μF, 50 V Solid Electrolytic Tantalum capacitors manufactured with conventional technology and F-Tech

Electrolytic Tantalum capacitors manufactured with F-Tech was also verified by the end-user in high-reliability application [104]. At the same time the failure rate is decreasing with time ($\beta < 1$) in Solid Electrolytic Tantalum capacitors manufactured with conventional technology and with F-Tech (Fig. 3.19b), which is typical for these capacitors within Weibull limitation $V_a/V_r \leq 1.527$ at 85 °C. The lower β in the capacitors manufactured with F-Tech indicates steeper failure rate decrease with time and, thereby, shorter additional time needed to achieve C-grade (0.01%/1000 h) or D-grade (0.001%/1000 h) failure rates required for the high-reliability applications.

The accelerated test of the H-case 220 µF, 25 V Polymer Tantalum capacitors manufactured with conventional technology and F-Tech was performed at several V_a/V_r and 105 °C. Figure 3.20 shows cumulative percent of failed capacitors as a function of time.

Figure 3.20a shows wear-out behavior in Polymer Tantalum capacitors manufactured with conventional technology: most of the parts fail within relatively short time intervals and these intervals are shifting to earlier times with increasing V_a/V_r. This performance is similar to shown by Paulsen et al. [99, 100] for Polymer Tantalum capacitors manufactured with conventional technology. In contrast to that, there is no wear-out in Polymer Tantalum capacitors manufactured with F-Tech (Fig. 3.20b) even at higher V_a/V_r and much longer time than these used for Polymer Tantalum capacitors manufactured with conventional technology.

Comparison of Figs. 3.19 and 3.20 shows early failures at the accelerated testing of Solid Electrolytic Tantalum capacitors while practically no early failures in Polymer Tantalum capacitors. The difference in the time-to-failure performance in these capacitors can be related to the difference between the high-temperature pyrolytic deposition of the MnO_2 cathode in Solid Electrolytic Tantalum capacitors and low-temperature deposition of the polymer cathode in Polymer Tantalum capacitors. The temperature accelerates degradation processes, particularly, growth of the crystalline inclusions in the amorphous matrix of the Ta_2O_5 dielectric, resulting in earlier failures in Solid Electrolytic Tantalum capacitors in comparison to that in Polymer Tantalum capacitors. By eliminating crystalline seeds at the dielectric-anode interface, F-Tech suppresses crystallization process in the Ta_2O_5 dielectric reducing failure rate in both types of solid tantalum capacitors and eliminating wear-out performance in Polymer Tantalum capacitors.

Using the data presented in Fig. 3.20b, the failure rate in Polymer Tantalum capacitors manufactured with F-Tech was estimated for 100 and 1000 h of accelerated testing (Fig. 3.21). The acceleration factor was calculated as $A = A_V \times A_T$, where A_V is voltage acceleration factor and A_T is temperature acceleration factor. According to the Weibull equation $A_V = 1824$ at $V_a/V_r = 1.4$ and according to the Arrhenius part of the Prokopovich-Vaskas equation $A_T = 11.09$ with $T_1 = 358$ K (85 °C), $T_2 = 378$ K (105 °C), and $E_a = 1.4$ eV as it was reported in [100] for given V_a/V_r.

As one can see in Fig. 3.21, Polymer Tantalum capacitors manufactured with F-Tech have the lowest D-grade failure rate, which is decreasing with time similar to that in Solid Electrolytic Tantalum capacitors.

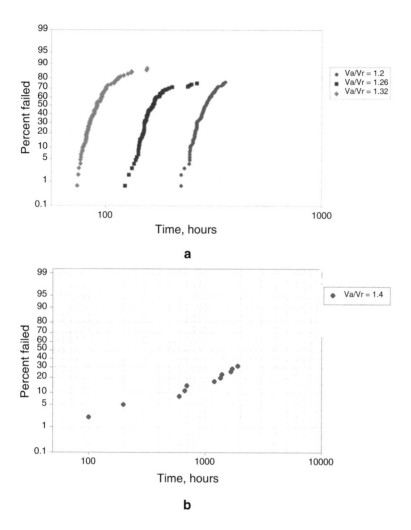

Fig. 3.20 Cumulative percent of failed parts versus time at accelerated test of H-case 220 μF, 25 V Polymer Tantalum capacitors manufactured with conventional technology (**a**) and F-Tech (**b**) at different V_a/V_r and 105 °C

Flawless dielectric technology using deox-sintered anodes (F-Tech with deox-sintering) allows to combine high reliability with record high volumetric efficiency in lower-voltage Polymer Tantalum capacitors [105]. As an example Fig. 3.22 shows cumulative percent of failed versus time and CV/cm^3 for D-case 220 μF, 16 V Polymer Tantalum capacitors manufactured with either conventional technology or F-Tech with deox-sintering and tested at $V_a/V_r = 1.3$ and 105 °C.

According to Fig. 3.22a, most of the Polymer Tantalum capacitors manufactured with conventional dielectric technology failed within relatively short time interval (wear-out region), while only 1 out of 200 Polymer Tantalum capacitors

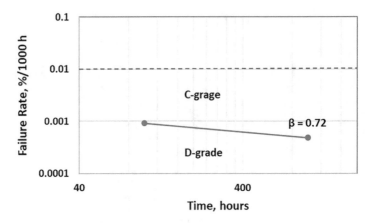

Fig. 3.21 Failure rate vs. time in H-case 220 μF, 25 V Polymer Tantalum capacitors manufactured with F-Tech and tested at $V_a/V_r = 1.4$ and 105 °C

manufactured with F-Tech with deox-sintering failed during this test. The high reliability of the Polymer Tantalum capacitors manufactured with F-Tech with deox-sintering is accompanied by about a 30% increase in CV/cm^3 in these capacitors in comparison with the CV/cm^3 in the Polymer Tantalum capacitors manufactured with conventional technology (Fig. 3.22b). Combining high reliability with high volumetric efficiency is unusual for tantalum capacitors. Typically high reliability is achieved with coarser powder anodes and thicker dielectrics, thus causing CV/cm^3 loss.

The characteristic failure mode in Solid Electrolytic and Polymer Tantalum capacitors manufactured with conventional technology and F-Tech is low insulation resistance or short. The failed short Solid Electrolytic Tantalum capacitors usually have black marks on their surface (Fig. 3.23) while typically there are no black marks on the surface of the failed short Polymer Tantalum capacitors.

According to Fig. 3.23, black marks on the surface of the failed short Solid Electrolytic Tantalum capacitors have smaller size in the capacitors manufactured with conventional technology (Fig. 3.23a) in comparison to the larger size black marks in the capacitors manufactured with F-Tech (Fig. 3.23b).

The fault sites in the dielectric were revealed in the failed short Solid Electrolytic and Polymer Tantalum after the epoxy compound and external layers of silver, carbon, and cathode were chemically stripped (Fig. 3.24).

As one can see in Fig. 3.24, in both Solid Electrolytic and Polymer Tantalum capacitors the fault sites have larger size in the failed capacitors manufactured with conventional technology (Fig. 3.24a, c) in comparison to the smaller size fault sites in the failed capacitors manufactured with F-Tech (Fig. 3.24b, d), which is in agreement with relatively large defect areas such as conglomerates of the crystalline inclusions in the dielectric in case of the conventional technology, while no such large defect areas in the dielectric in case of F-Tech.

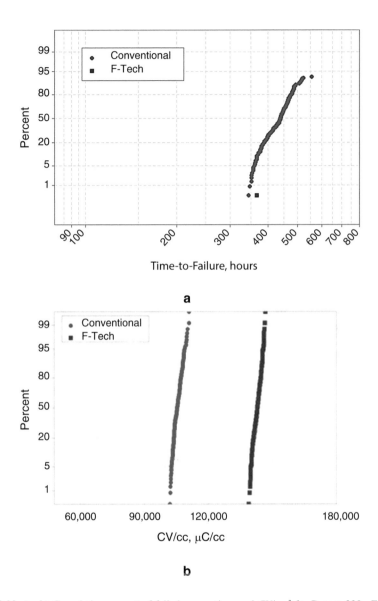

Fig. 3.22 (**a**, **b**) Cumulative percent of failed versus time and CV/cm^3 for D-case 220 μF, 16 V Polymer Tantalum capacitors manufactured with either conventional technology or F-Tech with deox-sintering and tested at $V_a/V_r = 1.3$ and 105 °C

The chemical composition inside and outside the fault sites in the dielectric in the failed short Solid Electrolytic Tantalum capacitors was investigated by energy dispersive X-ray EDX spectroscopy (Fig. 3.25).

As one can see in Fig. 3.25, the spectrums inside (Fig. 3.25a) and outside (Fig. 3.25b) the fault site in the dielectric have peaks of tantalum, oxygen and

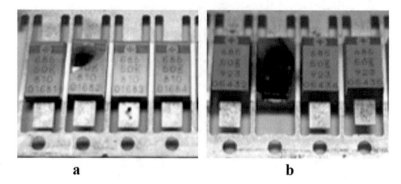

Fig. 3.23 Black marks on the surface of the failed short X-case 6.8 μF, 50 V Solid Electrolytic Tantalum capacitors with conventional technology (**a**) and F-Tech (**b**)

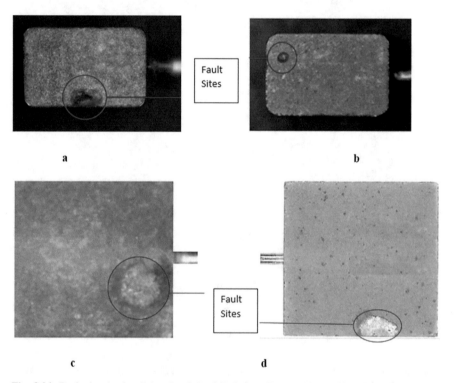

Fig. 3.24 Fault sites in the dielectric of the failed short X-case 6.8 μF, 50 V Solid Electrolytic Tantalum capacitors (**a, b**) and H-case 220 μF, 25 V Polymer Tantalum capacitors (**c, d**) manufactured with conventional technology (**a, c**) and F-Tech (**b, d**)

residual carbon and silicon from the epoxy compound and external layers. At the same time the spectrum inside the fault site in the dielectric has additional peaks of manganese. Since MnO_2 cathode was stripped from the dielectric, the peaks of manganese and oxygen in Fig. 3.25a indicate chemically stable low magnesium oxides:

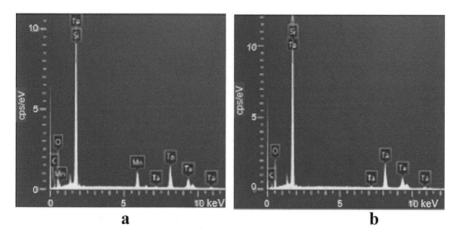

Fig. 3.25 EDX spectrums inside (**a**) and outside (**b**) the fault sites in the dielectric in the failed short X-case 6.8 μF, 50 V Solid Electrolytic Tantalum capacitor

Mn_2O_3, Mn_3O_4, and MnO that form on the surface of the fault site in the dielectric due to reduction of the MnO_2 cathode under the heat released in breakdown channel. The EDX spectroscopy was also performed inside and outside the fault sites in the dielectric of the failed short Polymer Tantalum capacitors. Both inside and outside spectrums were identical to shown in Fig. 3.25b with peaks of tantalum, oxygen, and residual silicon and carbon from the epoxy compound and the external layers on the surface of the dielectric.

Though oxygen was released from MnO_2 cathode when it was reduced into low manganese oxides at breakdown event and at the same time micropores were formed in the fault sites in the dielectric exposing tantalum anode, there were no signs of ignition and burning tantalum in failed short Solid Electrolytic Tantalum capacitors. Outside the small fault sites, tantalum anode was tightly covered with thick layer of undamaged anodic tantalum oxide (Fig. 3.26).

Principally different structure from the one shown in Fig. 3.26 was observed when porous tantalum anode was ignited with a burner (Fig. 3.27).

As one can see in Fig. 3.27, when tantalum anode is ignited, it emits intense white glow, indicating temperatures in excess of 1000 °C (Fig. 3.27b). At this temperature not only failed short tantalum capacitor, but also the surrounding electronic components and the circuit board would have been totally destroyed. However, no observations of such a dramatic damage have been reported for failed short Solid Electrolytic Tantalum capacitors. During the burning tantalum anodes convert into foamy polycrystalline tantalum pentoxide (Fig. 3.27c), and this type of the anode structure also has not been observed in any of the failed short SMD-type tantalum capacitors.

Presented in Figs. 3.25, 3.26, and 3.27 is evidence that low manganese oxides that form on the fault sites in the dielectric under the heat released at breakdown tightly seal the microcracks in the dielectric, preventing reaction between the

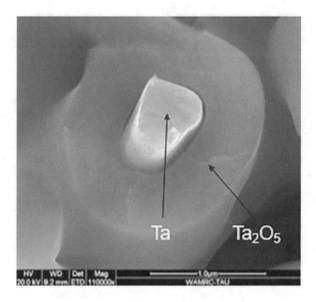

Fig. 3.26 Breakage of the tantalum anode in failed short X-case 6.8 μF, 50 V Solid Electrolytic
Tantalum capacitor

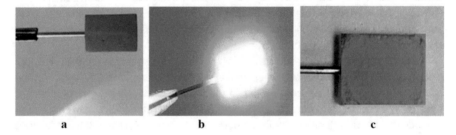

Fig. 3.27 Tantalum anode ignited with a burner (**a**), burning (**b**) and cooled to room temperature (**c**)

exposed tantalum anode and oxygen released from the reduced MnO_2 cathode and,
thereby, preventing ignition and burning tantalum in failed short Solid Electrolytic
Tantalum capacitors. The heat released at breakdown of the SMD-type Solid
Electrolytic and Polymer Tantalum capacitors propagates from the fault site in the
Ta_2O_5 dielectric through the epoxy compound toward the external surface of the
capacitor (Fig. 3.28).

For the same amount of energy stored in the capacitor, the amount of heat
released at breakdown is larger in failed short Solid Electrolytic Tantalum capacitors in comparison to that in Polymer Tantalum capacitors due to resistant low manganese oxides in the fault sites in the dielectric. As a result of the larger amount of
heat, failed short Solid Electrolytic Tantalum capacitors typically have black marks
on their external surface while typically there are no black marks on the surface of
the failed short Polymer Tantalum capacitors. Failed short Solid Electrolytic

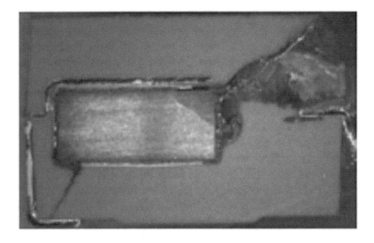

Fig. 3.28 Cross section of the failed short X-case 6.8 μF, 50 V Solid Electrolytic Tantalum capacitor

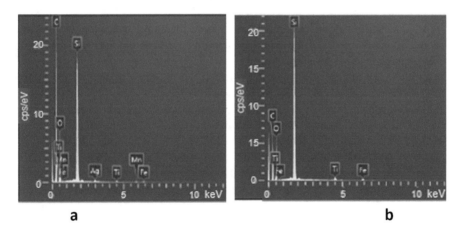

Fig. 3.29 EDX spectrums of the epoxy compound inside (**a**) and outside (**b**) the black mark on the surface of the epoxy compound in failed short X-case 6.8 μF, 50 V Solid Electrolytic Tantalum capacitor

Tantalum capacitors manufactured with F-Tech have smaller fault sites and, thereby, more heat released at breakdown, resulting in larger size black marks on their external surface in comparison to these in failed short Solid Electrolytic Tantalum capacitors manufactured with conventional technology (Figs. 3.23 and 3.24).

Chemical composition of the epoxy compound inside and outside the black marks on the surface of the failed short Solid Electrolytic Tantalum capacitor was investigated with EDX spectrometry (Fig. 3.29).

As one can see in Fig. 3.29, the major difference between the epoxy compound inside the black mark (Fig. 3.29a) and in undamaged yellow color epoxy compound outside the black mark (Fig. 3.29b) is high carbon content in the black mark indicating local carbonization of the epoxy compound under the heat released at breakdown. The heat induced transformations in the epoxy compound were investigated by the combined thermo-gravimetric analysis (TGA)—monitoring the weight changes with temperature under the controlled heat flow—and differential scanning calorimetry (DSC)—monitoring exothermic and endothermic events happening as a material is heated at a constant rate. The results of the TGA-DSC analysis of the epoxy compound in nitrogen and in air and the epoxy colors before and after the analysis are presented in Fig. 3.30.

According to Fig. 3.30a, the absence of air formation of the black carbonized epoxy is clearly observed. The chemical decomposition of the epoxy is initiated at about 350 °C and then carbonization of the materials occurs until 1000 °C as indicated by the continuous weight loss. The final weight loss is about 18%, which confirms carbon formation from epoxy. In the presence of air (Fig. 3.30b), the significant chemical decomposition of the epoxy starts at about 390 °C and then at about 450 °C the organic epoxy material is ignited and burned. The visual observation of the material left indicates that there is no carbon present. The TGA shows that total weight loss is about 25%. The remaining material is inorganic part of the compound, which does not change as material is heated from 600 to 1000 °C. Temperature-related transformations of the epoxy resins and composites including carbonization of the epoxy compound have been previously investigated in significant detail [106–109].

Comparison of Fig. 3.30a, b shows that in the presence of oxygen the epoxy network resists thermal oxidation up to about 400 °C, since the weight loss in nitrogen and air are practically the same in this temperature range. Carbonization of the epoxy material in vicinity of the fault site in the dielectric consumes heat released at breakdown and thus limits temperatures and protects from burning. The carbonization front is propagating to the external surface of the capacitor causing appearance of the black marks on the epoxy surface. Increasing the current in failed short Solid Electrolytic Tantalum capacitors can eventually increase the temperature up to the point where the epoxy compound ignites and burns; though, no ignition and burning are typically observed at normal conditions of the accelerated testing and field application. Similar ignition and burning of the epoxy compound can occur in any SMD-type electronic component that failed short and then subjected to unrestricted current. The epoxy compound used in these capacitors is flame retardant and meets UL 94 V-0 flammability standard requiring not more than 10 s of flaming after the burning is initiated.

In conclusion, there is strong impact of technology on reliability of Solid Electrolytic and Polymer Tantalum capacitors. Solid Electrolytic Tantalum capacitors manufactured with flawless technology (F-Tech) have significantly lower failure rate in comparison to the failure rate in the capacitors manufactured with conventional technology. Polymer Tantalum capacitors manufactured with F-Tech have the lowest failure rate, which is decreasing with time of the accelerated testing similar to that in Solid Electrolytic Tantalum capacitors. At the same time, Polymer Tantalum capacitors manufactured with conventional technology typically have no

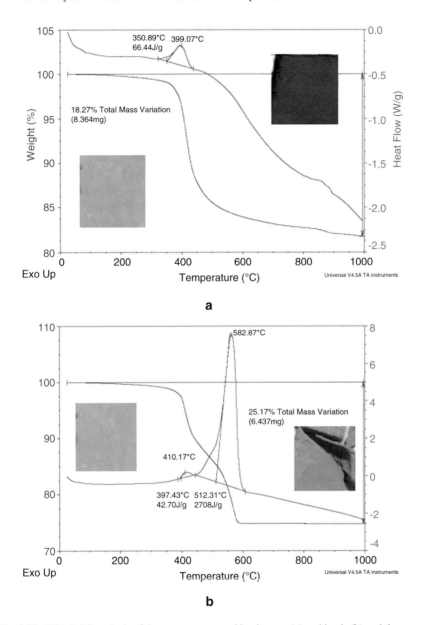

Fig. 3.30 TGA-DSC analysis of the epoxy compound in nitrogen (**a**) and in air (**b**) and the epoxy color before and after the analysis

or very few failures at the initial phase of the testing and then majority of the parts fail within short time interval (wear-out region). Lower-voltage Polymer Tantalum capacitors manufactured with F-Tech with deox-sintering combine high reliability with record high volumetric efficiency.

There is no ignition and burning tantalum in failed short SMD-type Solid Electrolytic Tantalum capacitors. The low manganese oxides, which form on the surface of the fault sites in the dielectric under the heat released at breakdown, seal microcracks in the dielectric, preventing the reaction between the exposed tantalum anode and the oxygen released from reduced MnO_2 cathode and, thereby, eliminating ignition and burning tantalum. The black marks are typically observed on the surface of the failed short Solid Electrolytic Tantalum capacitors while typically there are no black marks on the surface of the failed short Polymer Tantalum capacitors. These black marks are the areas of the epoxy compound carbonized under the heat propagated from the fault sites in the dielectric at the breakdown event. For the same amount of energy stored in the capacitor, more heat is generated in failed short Solid Electrolytic Tantalum capacitors in comparison to Polymer Tantalum capacitors due to the high resistance low manganese oxides covering the fault sites in the dielectric. Small size of the fault sites in the dielectric of the capacitors manufactured with F-Tech also causes more heat released at breakdown and, thereby, larger size of the black marks.

No ignition and burning of the epoxy compound are typically observed when SMD-type Solid Electrolytic and Polymer Tantalum capacitors fail short at normal conditions of the accelerated testing or during field applications. The epoxy compounds used in manufacturing of SMD-type tantalum capacitors and other types of the electronic components are flame retardant and meet high safety standard.

3.3   High-Voltage Polymer Tantalum Capacitors

The original Polymer Tantalum capacitors with a poly(3,4-ethylenedioxythiophene) (PEDOT) cathode had low BDV that increased to approximately 50 V and then leveled off as the dielectric thickness continued to grow. This limitation together with high and unstable DCL has made applications of original Polymer Tantalum capacitors impossible in high-voltage and high-reliability circuits, where low ESR and high ripple current capability are most critical. Due to the major changes in technology present Polymer Tantalum capacitors are characterized by the highest working voltage and lowest DCL ever achieved in solid tantalum capacitors (Fig. 3.31) [105].

The important change to the technology that contributed to better performance and reliability of Polymer Tantalum capacitors relates to the process of applying a PEDOT cathode on the surface of the Ta_2O_5 dielectric. Traditionally it was an in situ chemical reaction between the monomer and the oxidant (in situ PEDOT) followed by washing out by-products of the chemical reaction. The new process consists of dipping previously formed tantalum anodes in a water-based pre-polymerized PEDOT dispersion (slurry PEDOT) followed by drying in air at both room temperature and elevated temperature. The slurry PEDOT process does not have any by-products that contaminate the polymer cathode and its interface with the Ta_2O_5 dielectric. At the same time, it is practically impossible to fully wash out by-products of the in situ chemical reaction from the small pores in sintered and formed

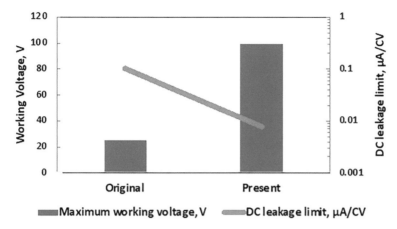

Fig. 3.31 Maximum working voltage and DC leakage per CV limit in Polymer Tantalum capacitors

tantalum anodes, especially when they become filled with the polymer cathode. Therefore, the absence of by-products is a primary advantage of the slurry PEDOT process. As an example, Fig. 3.32 presents fragments of the second ion mass spectrometry (SIMS) spectra of in situ PEDOT and slurry PEDOT inside porous anodes of Polymer Tantalum capacitors.

According to Fig. 3.32, there is a strong peak of residual iron with ion mass over charge ratio 55.93 on in situ PEDOT spectra (Fig. 3.32a) and no iron peak on the slurry PEDOT spectra (Fig. 3.32b). The source of iron is oxidant solution of iron (III) p-toluenesulfonate (FePTS) used in the chemical reaction on in situ PEDOT.

A combination of flawless technology (F-Tech) in the anode manufacturing, which eliminates defects such as micropores, cracks, and crystalline inclusions in the Ta_2O_5 dielectric, and slurry PEDOT in the cathode manufacturing resulted in a radical change to I–V characteristics of Polymer Tantalum capacitors [81, 110]. As an example, Fig. 3.33 shows I–V characteristics of Polymer Tantalum capacitors with F-Tech anodes, 200 nm anodic oxide films, and either pre-polymerized slurry PEDOT or in situ PEDOT cathodes.

As one can see in Fig. 3.33, Polymer Tantalum capacitors with F-Tech anodes and slurry PEDOT cathodes have nearly ideal diode-like I–V characteristics with very low current at normal polarity (+ on tantalum anode) and rapid current increase at reverse polarity (− on tantalum anode). In comparison, Polymer Tantalum capacitors with in situ PEDOT have significantly higher current at normal polarity with slight difference in current at the reverse polarity.

These I–V characteristics were explained in [81, 110] within classical MIS theory. In this case, M stands for tantalum anode, I stands for Ta_2O_5 insulator, and S stands for p-type semiconductor PEDOT cathode. According to this theory broadly used in physics of semiconductor devices, current flow through the $Ta/Ta_2O_5/$ PEDOT structure at normal polarity can be caused by field- and temperature-induced emission of current carriers from the semiconductor into the insulator. A potential barrier at the $Ta_2O_5/$PEDOT interface limits the emission. This barrier

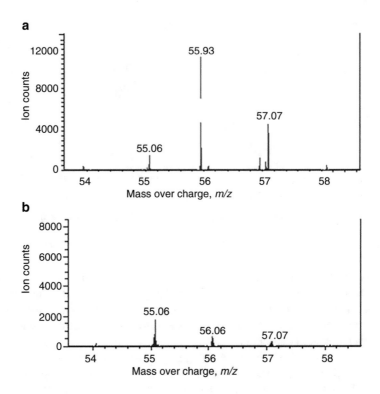

Fig. 3.32 SIMS spectra of in situ PEDOT (**a**) and slurry PEDOT (**b**) inside porous anodes of Polymer Tantalum capacitors

Fig. 3.33 *I–V* characteristics of Polymer Tantalum capacitors with F-Tech anodes and either slurry PEDOT or in situ PEDOT cathodes

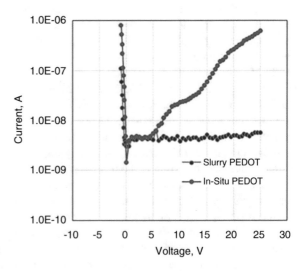

effectively increases with applied voltage, which keeps the current low and nearly constant in the case of flawless dielectric provided by F-Tech, while pores and cracks shunt the dielectric in case of conventional technology and impair the potential barrier on the Ta_2O_5/PEDOT interface.

In the case of an in situ PEDOT cathode, residuals of the chemical reaction between the oxidant and monomer cause surface charge at the Ta_2O_5/PEDOT interface. This surface charge "pins" the potential barrier at the IS interface and prevents its increase with applied voltage at normal polarity, resulting in higher current and lower BDV at normal polarity. Increasing the number of in situ polymerization cycles to improve the coverage of the Ta_2O_5 dielectric with polymer cathode makes it more difficult to wash out by-products of the in situ chemical reaction from the small pores in tantalum anodes. The increased contamination of the polymer cathode with by-products of the in situ chemical reaction results in a higher-density surface charge and even higher DCL and lower BDV.

The combination of F-Tech anodes, special formation conditions [67, 68] that suppress field crystallization of the amorphous matrix of the Ta_2O_5 dielectric, and slurry PEDOT cathodes provides record high BDV in solid tantalum capacitors. Figure 3.34 demonstrates BDV dependence on formation voltage in tantalum capacitors with F-Tech anode and either slurry PEDOT, in situ PEDOT, or MnO_2 cathode [110–112].

As one can see in Fig. 3.34, the ultimate BDV in tantalum capacitors with F-Tech anodes and slurry PEDOT cathodes is approaching 250 V, which allows increasing working voltage in these capacitors up to 125 V never seen before in any type of solid tantalum capacitors.

Even at the highest working voltages, Polymer Tantalum capacitors with F-Tech anodes and slurry PEDOT cathodes demonstrate a low and stable DCL comparable to the DCL in Wet Tantalum capacitors. As an example, Fig. 3.35 presents DCL distributions in B-case 100 μF–60 V Polymer Hermetic Seal (PHS) Tantalum capacitors at different durations of the life test at rated voltage and 85 °C [113].

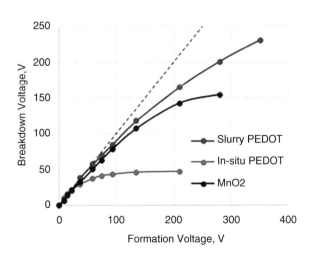

Fig. 3.34 BDV vs. formation voltage in tantalum capacitors with F-Tech anode and either slurry PEDOT, in situ PEDOT, or MnO_2 cathode

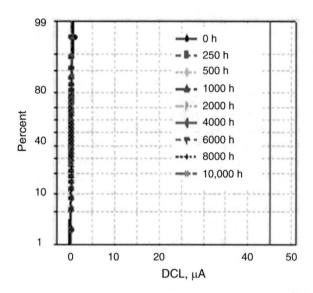

Fig. 3.35 DCL distributions in B-case 100 µF–60 V PHS Tantalum capacitors during life test at rated voltage and 85 °C

As one can see in Fig. 3.35, all the DCL distributions from the initial readings (0 h) up to the final readings (10,000 h) of the life test overlap each other, demonstrating exceptional DCL stability at the long-term testing without de-rating. The total lack of failures during the 10,000 h life test was particularly due to the simulated breakdown screening (SBDS) included in the manufacture process. No failures were also registered during short-term electrical and mechanical tests included in the qualification protocol. These results allowed usage of PHS Tantalum capacitors in the most critical high-voltage and high-reliability applications such as military and space.

While DC leakage and maximum working voltage are comparable in Wet and PHS Tantalum capacitors with F-Tech anodes and slurry PEDOT cathodes, their AC characteristics are very different. As an example, Fig. 3.36 presents ESR dependence on frequency in the range of operating temperatures in B-case 82 µF–75 V Wet and PHS Tantalum capacitors [114].

As one can see in Fig. 3.36, at frequency 100 kHz typically used to test ESR in tantalum capacitors, ESR in Wet Tantalum capacitor is relatively low at elevated temperatures but increases sharply at lower temperatures (Fig. 3.36a). In contrast, ESR in Polymer Tantalum capacitor remains low in the entire range of operating temperatures (Fig. 3.36b). The major contributor to the high frequency ESR in these capacitors is the resistance of their cathodes. In Wet Tantalum capacitors, resistance of the liquid electrolyte cathode depends on the concentration and mobility of ions in the electrolyte. When ambient temperature approaches the freezing temperature of the electrolyte, mobility of the ions becomes negligible, causing a sharp increase in the resistance of the cathode and, thereby, ESR in Wet Tantalum capacitors.

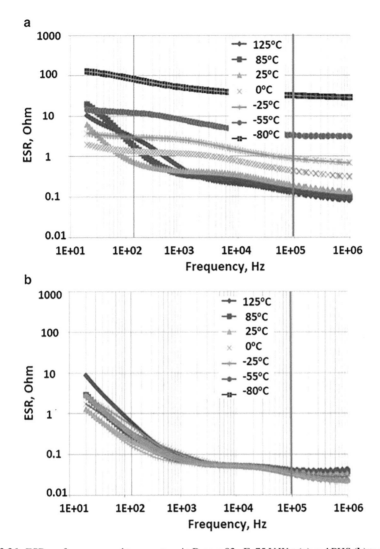

Fig. 3.36 ESR vs. frequency and temperature in B-case 82 μF–75 V Wet (**a**) and PHS (**b**) tantalum capacitors

At the same time, resistance of the cathode in Polymer Tantalum capacitor is low and practically independent on temperature due to the high concentration and high mobility of p-type current carriers in PEDOT cathode doped with poly(styrene sulfonate) (PSS) acid. Depending on the level of doping, the PEDOT-PSS system demonstrates properties of a solid organic semiconductor with narrow band gap or even degenerated semiconductor [115].

The differences in ESR in Wet and Polymer Tantalum capacitors determine the differences in their capacitance dependence on frequency and temperature. As an example, Fig. 3.37 presents capacitance dependence on frequency in the range of operating temperatures for B-case 82 μF–75 V Wet and PHS Tantalum capacitors [114].

As one can see in Fig. 3.37, Wet and Polymer Tantalum capacitors have almost equal capacitance at 120 Hz and room temperature typically used to test capacitance in tantalum capacitors. The capacitance eventually decreases at higher frequencies and lower temperatures; however, it remains stable in a much broader range of frequencies and operating temperatures in Polymer Tantalum capacitor (Fig. 3.37b) in

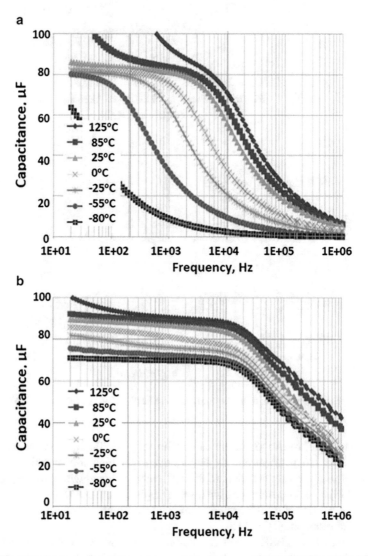

Fig. 3.37 Capacitance vs. frequency and temperature in B-case 82 μF–75 V Wet (**a**) and PHS (**b**) Tantalum capacitors

comparison to Wet Tantalum capacitor (Fig. 3.37a). J. Prymak explained the capacitance loss with frequency in tantalum capacitors by RC ladder—the longer path for electric signal and, thereby, higher resistance of the cathode in the core of the porous tantalum anodes in comparison to their external surface [101]. At some initial frequency, the electric signal can't penetrate the anode core resulting in initial capacitance loss, which progresses at higher frequencies. For the capacitors with equal size and porosity of the anodes, capacitance loss with frequency is much more pronounced in Wet Tantalum capacitors than in Polymer Tantalum capacitors due to the differences in resistivity of their liquid electrolyte and conductive polymer cathodes, especially, at low operating temperatures.

In the lower-voltage Polymer Tantalum capacitors manufactured with higher CV/g tantalum anodes, internal impregnation of the finer pores in formed tantalum anodes with slurry PEDOT particles becomes insufficient. In this case, a hybrid technology is typically used, combining in situ PEDOT technology for internal impregnation and slurry PEDOT technology for external coverage of the tantalum anodes. In some low-voltage high CV Polymer Tantalum capacitors, in situ PEDOT technology is used for both internal impregnation and external coverage of the formed tantalum anodes to achieve minimum ESR in these capacitors.

In conclusion, the combination of F-Tech anode and slurry PEDOT cathode allows for manufacturing of Polymer Tantalum capacitors with record high BDV and working voltage and record low DCL. The anodizing conditions, which suppress field crystallization of the Ta_2O_5 dielectric during its formation, and low-temperature deposition of polymer cathode in comparison to pyrolytic deposition of MnO_2 cathode, which suppresses thermal crystallization of the formed Ta_2O_5 dielectric, also contribute to the high BDV, low DCL, and high reliability of these capacitors. The MIS model explained the unique DC performance of Polymer Tantalum capacitors with F-Tech anode and slurry PEDOT cathode by potential barrier on the Ta_2O_5-PEDOT interface that efficiently limits DC current through the capacitor. At the same time, in Polymer Tantalum capacitors with traditional anode and in situ PEDOT cathode technologies, this barrier is inefficient because it's pinned by surface charge from by-products of the in situ chemical reaction and shunted by the defects in the dielectric. High-voltage Polymer Tantalum capacitors also demonstrate low ESR and high capacitance and ESR stability with frequency and temperature due to low and stable temperature resistivity of the polymer cathode in these capacitors.

3.4 Asymmetric Conduction and Stability of DC Characteristics

Wet Tantalum capacitors with tantalum anode, anodic oxide of tantalum as a dielectric, and liquid electrolyte cathode demonstrate asymmetric conduction with low DC leakage and high breakdown voltage at normal polarity (+ on tantalum anode) and high DCL and low BDV at reverse polarity (− on tantalum anode) [17]. The group of metals with this type of asymmetric conduction of their anodic oxide films in liquid cells, known as valve metals, also includes Al, W, Ti, Hf, Nb, and Zr [17]. Solid Electrolytic Tantalum capacitors with manganese dioxide cathode and Polymer Tantalum capacitors with inherently conducting polymer cathode also demonstrate asymmetric conduction [5, 116]. Metal-Oxide-Metal (MOM) Tantalum capacitors with vacuum sputtered tantalum film as an anode, anodic oxide film of tantalum as the dielectric, and vacuum sputtered metal counter electrode typically also demonstrate asymmetric conduction [19, 117].

Despite the commonly demonstrated asymmetric conduction in Wet, Solid Electrolytic Polymer, and MOM Tantalum capacitors, the degree of the asymmetry varies significantly depending on the materials and processes used in manufacturing of these capacitors. Particularly in MOM Tantalum capacitors, conduction asymmetry diminishes with smaller area electrodes and thinner Ta_2O_5 dielectric, and ultimately conduction becomes symmetric [19, 118]. On the other hand, heat treatment of the tantalum anode with a Ta_2O_5 dielectric formed on its surface, as well as an increase in the dielectric thickness, results in stronger conduction asymmetry in these capacitors [19, 116, 119].

Test conditions also have a significant effect on the conduction asymmetry in tantalum capacitors. At liquid nitrogen temperature, Polymer Tantalum capacitors demonstrate symmetric conduction with negligible current at both polarities even at elevated voltages [81]. However, conduction asymmetry in these capacitors was observed to increase at higher temperatures and higher applied voltages. Similar results were presented for tantalum capacitors with MnO_2 cathodes [120, 121].

The presence of moisture also plays an important role in the conduction asymmetry in solid tantalum capacitors. In MOM Tantalum capacitors with gold counter electrodes, BDV at normal polarity was higher in the presence of moisture, while BDV at reverse polarity was not affected by moisture [116]. When the gold counter electrode was replaced with a liquid electrolyte, strong conduction asymmetry was observed with high current and low BDV at reverse polarity.

The DCL at normal and reverse polarities in tantalum capacitors can change with time at constant temperature, applied voltage, and ambient conditions, affecting conduction asymmetry in these capacitors. A reverse DCL increase with time at relatively low reverse voltages and temperatures was demonstrated for solid tantalum capacitors with MnO_2 cathodes [120, 121]. The rate of the reverse DCL increase with time was higher at higher temperature and higher magnitude of the reverse voltages. Under these conditions reverse DCL could rapidly increase 3–4 orders of magnitude in comparison to the initial DCL at the beginning of the test. In some solid tantalum capacitors, reverse DCL could be restored to its initial level by applying voltage at normal polarity for some period of time, while in other capacitors, permanent damage to the dielectric was detected.

Numerous mechanisms have been presented in the literature to explain conduction asymmetry. These mechanisms can be roughly divided into two major groups: heterogeneous mechanisms focusing on local current flow through the structural flaws in the dielectric [17, 118] and homogeneous mechanisms assuming uniform current flow through the dielectric and focusing on the band structure of the dielectric and potential barriers at the dielectric interfaces with the electrodes [119, 122]. In some cases, a combination of the heterogeneous and homogeneous mechanisms was considered [117, 123].

It is obvious that there is no universal mechanism that can explain all the variety of experimental data on conduction asymmetry in tantalum capacitors made with different materials and processes and tested under different conditions. The detail analysis of the conduction asymmetry and DCL changes with time was performed on Polymer Hermetic Seal (PHS) Tantalum capacitors [124]. Hermetic design of the PHS Tantalum capacitors allows for a varying atmosphere inside the capacitor, in particular the amount of moisture, in a controlled manner as well as maintaining this atmosphere during the long-term testing. These conditions are important in order to determine the dominant mechanisms responsible for the DCL degradation and failures in Polymer Tantalum capacitors at normal and reverse polarities.

For analysis of the conduction asymmetry and DCL stability with time in [124], PHS Tantalum capacitors were manufactured with F-tech tantalum anodes sintered with 12,000 μC/g tantalum powder. The tantalum anodes were formed in an aqueous solution of polyethylene glycol and phosphoric acid using formation voltages 150 V for finished capacitors with a working voltage 60 and 190 V for working voltage 75 V. As it was described earlier, a pre-polymerized slurry PEDOT cathode with PSS dopant was applied by dipping the formed tantalum anodes into a waterborne dispersion of the nanoscale PEDOT-PSS particles and subsequent drying in air. The capacitors with external layers of carbon and silver were soldered inside brass cans and then hermetically sealed after drying at 125 °C for 24 h (dry) or after humidification at room temperature and 85% RH for 24 h (humid).

The current-voltage measurements in [124] were performed on these capacitors in broad range of temperatures. The current was allowed to decay for 300 s at constant voltage to reach a steady-state condition before its final value was recorded. The long-term DCL stability at normal polarity was investigated on a group of at least 12 capacitors through a 2000 h life test at working voltage and 85 °C, with intermediate measurements taken at 250, 500, and 1000 h of the life test. At reverse polarity DCL was monitored at constant voltage and I-time measurements were recorded. Breakdown voltage was tested at room temperature on a group of ten capacitors with a 1 A fuse in series with each capacitor. Voltage was increased at a constant rate of 1 V/min from 0 V to the voltage at which the fuse opened, which was defined as a breakdown event.

A typical asymmetric I–V characteristic of an as-manufactured B-case 75 μF–75 V PHS Tantalum capacitor is presented in Fig. 3.38 for normal and reverse polarities with a temperature range from room temperature to 85 °C.

Similar to previous results [81, 113], the current at normal polarity was very low and practically flat with voltage with a weak dependence on temperature, while the current at reverse polarity increased sharply with the magnitude of reverse voltage and temperature. In the given voltage and temperature ranges, the I–V characteristics of the as-manufactured Ta capacitors were practically identical for both the humid and dry capacitors. However, BDV and DCL stability with time at normal

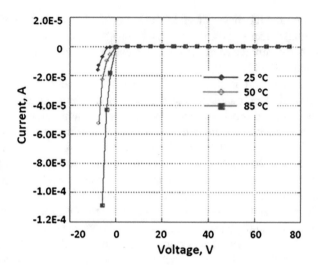

Fig. 3.38 *I–V* characteristics of the as-manufactured B-case 75 μF–75 V PHS Tantalum capacitor at normal (Tantalum +) and reverse (Tantalum −) polarities

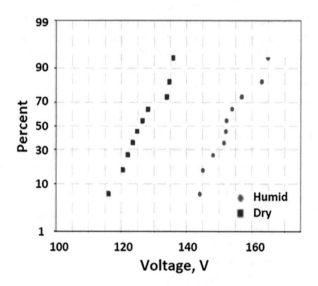

Fig. 3.39 BDV in B-case 75 μF–75 V humid and dry PHS Tantalum capacitors

and reverse polarities were significantly different in these capacitors. The BDV at normal polarity in B-case 75 μF–75 V humid and dry PHS Tantalum capacitors is presented in Fig. 3.39.

Higher BDV in the humid PHS Tantalum capacitors is correlated with higher DCL stability with time during life test at rated voltage and 85 °C in comparison to dry PHS Tantalum capacitors (Fig. 3.40).

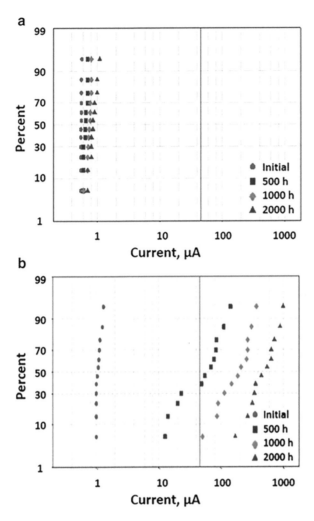

Fig. 3.40 DCL during life test at rated voltage and 85 °C in B-case 75 µF–75 V humid (**a**) and dry (**b**) PHS Tantalum capacitors

As shown in Fig. 3.40a, there was slight change in DCL in humid PHS Tantalum capacitors during the life test, while DCL in dry PHS Tantalum capacitors was observed to continuously increase during the life test, eventually resulting in parametric failures in these capacitors, as shown in Fig. 3.40b. Comparison of Figs. 3.39 and 3.40 shows that the presence of moisture inside the 75 V PHS Tantalum capacitors is crucial for high BDV and long-term DCL stability in these capacitors. Similar behavior was detected in 60 V PHS Ta capacitors during the BDV test and life test at rated voltage and 85 °C. The BDV at normal polarity in B-case 100 µF–60 V humid and dry PHS Tantalum capacitors is presented in Fig. 3.41.

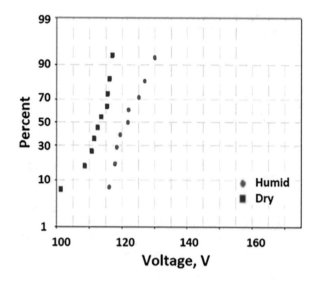

Fig. 3.41 BDV in humid and dry B-case 100 µF–60 V PHS Tantalum capacitors

From Fig. 3.41, humid 60 V PHS Tantalum capacitors had higher BDV in comparison to BDV in dry 60 V PHS Tantalum capacitors. DCL in humid 60 V capacitors was practically unchanged during the life test, as shown in Fig. 3.42a. However, DCL in dry 60 V capacitors was observed to increase significantly during this test as shown in Fig. 3.42b, although no failures were detected in this case.

Comparison of BDV and DCL behavior in PHS Tantalum capacitors with thicker dielectric (Figs. 3.39 and 3.40) and thinner dielectric (Figs. 3.41 and 3.42) shows that the beneficial effects of moisture are not as pronounced in the thinner dielectrics as they are in the thicker dielectrics; however, the presence of moisture is still quite crucial for the performance and reliability of both types of capacitors at normal polarity. Thus, controlled amounts of moisture are added to the PHS Tantalum capacitors before they are hermetically sealed [125]. As shown earlier in Fig. 3.35, very low and stable DCL readings during 10,000 h of life test at rated voltage and 85 °C were obtained on PHS Tantalum capacitors with controlled amount of moisture inside the hermetic can.

DCL behavior of humid and dry PHS Tantalum capacitors at reverse polarity is quite different from that at normal polarity. Figure 3.43 presents the reverse DCL versus time at 70 °C and reverse voltage −1 V in several humid B-case 75 µF–75 V PHS Tantalum capacitors.

According to Fig. 3.43, there are three stages of reverse DCL evolution during the reverse voltage test. Stage 1 represents low and stable reverse DCL during the first 2 h of the measurement. Stage 2 represents slow reverse DCL increase with some "noise" during 2–6 h of the measurement. Finally, stage 3 represents a rapid increase in reverse DCL and highly erratic behavior during 6–8 h of the reverse current measurement. Similar DCL behavior during a reverse bias measurement has

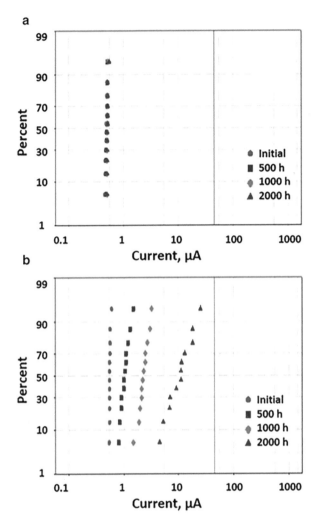

Fig. 3.42 DCL during life test at rated voltage and 85 °C in humid (**a**) and dry (**b**) B-case 100 μF–60 V PHS Tantalum capacitors

been observed previously for solid tantalum capacitors with MnO_2 cathodes [121–123].

Several of the 75 V PHS Tantalum capacitors were removed from the reverse bias measurement at the end of the stages 1, 2, and 3, respectively. They were subjected to 125 h conditioning process at rated voltage and 85 °C, cooled down to room temperature, and characterized again at rated voltage 75 V and normal polarity. In Fig. 3.44, the DCL readings at normal polarity prior to the reverse bias measurement, as well as after the three stages of the reverse bias measurement and subsequent conditioning at normal polarity, are presented.

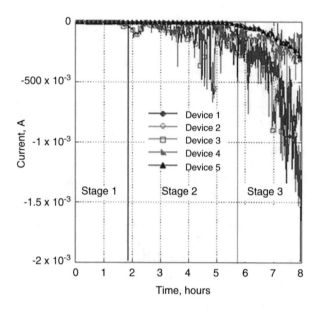

Fig. 3.43 DCL versus time at reverse voltage −1 V and 70 °C in humid B-case 75 μF–75 V PHS Tantalum capacitors

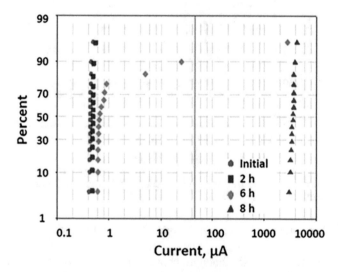

Fig. 3.44 DCL at normal polarity in humid B-case 75 μF–75 V PHS Tantalum capacitors before and after reverse bias measurement and conditioning at normal polarity

According to Fig. 3.44, DCL is fully recovered to its initial level after stage 1 (2 h) of the reverse voltage test. After stage 2 (6 h), one of the capacitors failed short, and DCL in two other capacitors was elevated, while DCL in the other capacitors recovered close to the initial DCL level. This was despite the fact that the reverse current increased several orders of magnitude during the 6 h of the reverse voltage measurement. After stage 3 (8 h), all of the capacitors failed short. It is clear that only stage 1, with low and stable reverse DCL, can be considered safe for applications of the solid tantalum capacitors at reverse polarity.

Typical reverse DCL versus time for the dry B-case 75 µF–75 V PHS Tantalum capacitors measured at 70 °C and −1 V is presented in Fig. 3.45.

From this graph, it is clearly shown that DCL in dry 75 V PHS Tantalum capacitors was very low and stable during the 8 h reverse voltage measurement. No change in DCL at 75 V normal polarity was observed after the 8 h reverse voltage measurement and 125 h of conditioning at rated voltage and 85 °C, as shown in Fig. 3.46.

DCL behavior of humid and dry 60 V PHS Tantalum capacitors at reverse polarity is qualitatively similar to that of 75 V PHS Ta capacitors. Figure 3.47 presents the reverse DCL versus time at 70 °C and reverse voltage −1 V in several humid B-case 100 µF–60 V PHS Tantalum capacitors.

Similar to the results shown in Fig. 3.43 for the 75 V humid PHS Tantalum capacitors, reverse DCL in the 60 V humid PHS Tantalum capacitors was initially low and stable but then eventually became erratic and gradually increased with time. After completion of the 8 h reverse voltage measurement, the capacitors were subjected to the 125 h conditioning at rated voltage and 85 °C, cooled down to room

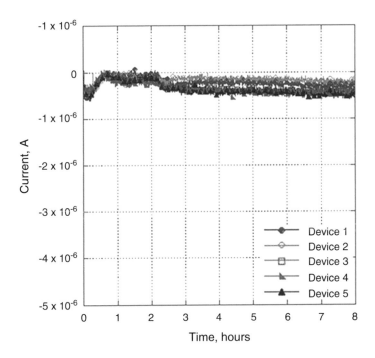

Fig. 3.45 DCL versus time at reverse voltage −1 V and 70 °C in dry 75 µF–75 V PHS Tantalum capacitors

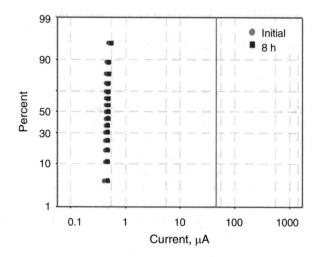

Fig. 3.46 DCL at normal polarity in dry 75 μF–75 V PHS Tantalum capacitors before and after reverse bias measurement and conditioning at normal polarity

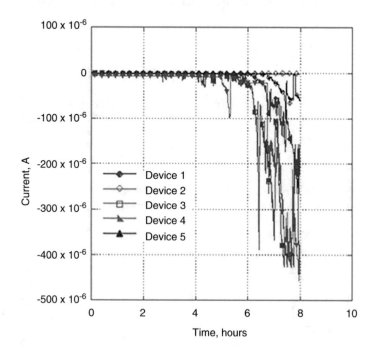

Fig. 3.47 DCL versus time in humid 100 μF–60 V PHS Tantalum capacitors at a reverse voltage of −1 V and 70 °C

temperature, and measured again at 60 V normal polarity. The DCL at normal polarity before and after the 8 h reverse voltage measurement and subsequent conditioning at normal polarity is shown in Fig. 3.48.

From Fig. 3.48, little change in DCL was observed for most of the 60 V humid PHS Tantalum capacitors as a result of the 8 h reverse voltage measurement, although in some of the capacitors the DCL was elevated. No shorts or parametric failures were observed. Comparison of the reverse voltage measurements of 75 and 60 V humid PHS Tantalum capacitors indicates that lower-voltage capacitors with thinner dielectrics demonstrate better reverse DCL stability and less change in normal polarity DCL as a result of the reverse voltage application. Similar to the results shown in Figs. 3.45 and 3.46 for the 75 V capacitors, reverse DCL in the dry 60 V PHS Tantalum capacitors was very low and stable and did not cause any changes in DCL at normal polarity.

These experimental results demonstrate that the presence of moisture has a strong effect on DCL stability in Polymer Tantalum capacitors at normal and reverse polarities. Moisture helps stabilize DCL at normal polarity while causing DCL instability at reverse polarity. Comparison of the DCL behavior in 75 and 60 V PHS Tantalum capacitors shows that lower-voltage capacitors with thinner dielectrics are more stable and less sensitive to the effects of moisture at both normal and reverse polarities. As it was presented earlier, thicker amorphous dielectrics are more prone to crystallization because the increase in internal energy due to disordering of their bulk is not compensated by the decrease in internal energy due to the absence of misfit dislocations at their interface with the crystalline substrate. Coarser tantalum powders sintered in vacuum at higher temperatures usually have higher chemical purity and smoother morphology in comparison to finer tantalum powders sintered at lower temperatures. Larger pores in these anodes allow for better coverage of the

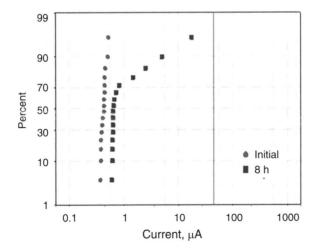

Fig. 3.48 DCL at normal polarity in humid 100 μF–60 V PHS Tantalum capacitors before and after the 8 h reverse voltage measurement and conditioning at normal polarity

Ta$_2$O$_5$ dielectric with conductive polymer cathode. It is obvious that for the same dielectric thickness and CV/g of the tantalum powder, the probability of having structural defects in the dielectric is reduced in smaller size anodes and in anodes sintered with flawless technology (F-Tech).

The effect of moisture on DCL in Polymer Tantalum capacitors is clearly polarity dependent. As it was presented in [81, 110, 124], at normal polarity residual moisture works as a plasticizer, maintaining a sufficient potential barrier at the Ta$_2$O$_5$/p-type PEDOT interface and providing high BDV and low and stable DCL in humid Polymer Tantalum capacitors. The effect of the plasticizer is primarily achieved through an increase in polymer chain mobility due to addition of small molecules that dissolve in the polymer, separating the chains from each other and hence making the polymer chain movement easier. It was determined that the pre-polymerized slurry PDEOT cathode absorbed an amount of water equal to about 8% of the dry polymer weight when exposed to the air with a relative humidity of about 70% for several hours. In presence of humidity, the barrier at the Ta$_2$O$_5$/polymer interface remains intact practically indefinitely maintaining low and stable DCL during life testing and field application. In dry Polymer Tantalum capacitors, a small amount of residual moisture can redistribute, leaving the barrier without a plasticizer. This results in an increase in DCL, which can be nonuniform, causing local overheating, oxygen migration and crystalline inclusion growth, and, thereby, further increase in DCL. Larger size and density of the structural defects in the dielectric intensify these degradation processes.

The effect of the plasticizer is also achieved by improving self-healing properties in Polymer Tantalum capacitors with slurry PEDOT-PSS cathode. In this case, self-healing can be achieved by local separation between the PEDOT and dopant PSS molecules in vicinity of the defect sites in the dielectric with high current density and rising temperature (de-doping) [81]. The molecules of separated PEDOT and PSS have higher resistivity in comparison to resistivity of the PEDOT-PSS, blocking current flow through the defect sites in the dielectric and preventing thermal runaway.

At reverse polarity different mechanisms can dominate effects of moisture on Polymer Tantalum capacitors. In humid capacitors with initially low leakage as observed in stage 1 of Fig. 3.43, the H$_2$O molecules can dissociate, and the H$^+$ ions (protons) will accelerate and diffuse through the dielectric, forming positive charge in the dielectric in the vicinity of tantalum anode, now effectively acting as a cathode. This charge layer can act to reduce the potential barrier at the Ta/Ta$_2$O$_5$ interface and locally increase the electrical field in the dielectric, resulting in higher current injection in the dielectric even at low reverse voltage especially in areas with rougher anode morphology and crystalline seeds in the dielectric [110]. These processes lead to the leakage current increase with time as observed on stage 2 on Fig. 3.43. Applying rated voltage at normal polarity after this stage of the reverse voltage test allows to extract positive charge from the dielectric and restore low DCL in most of the capacitors. However, continuing application of the reverse voltage will lead to further increase in current, local overheating with unrepairable damage of the dielectric, and eventually shorts as observed in stage 3 of Fig. 3.43. In dry tantalum capacitors under reverse bias, the production of H^+ ions is very unlikely to occur and leakage current remains low and stable for long period of time. The controlled amount of moisture added to the PHS Tantalum capacitors prior to their sealing in the can [125] provides long-term stability and reliability at normal polarity and also allows some reverse bias capability.

The presence of moisture in hermetic and non-hermetic Polymer Tantalum capacitors can cause failures common for many types of the electronic components such as corrosion, the popcorn effect, delamination, ion migration, tin whisker growth, etc. As an example, Fig. 3.49 shows popcorn effect at assembly (a), delamination of the carbon and silver external layers (b), migration of silver from the external silver layer toward the Ta_2O_5 dielectric (c), and tin whisker growth from pure tin solder (d) during long-term life test and storage of PHS Tantalum capacitors with humidity inside the hermetic can.

To eliminate moisture-related failures in PHS Tantalum capacitors, special technologies and materials are implemented such as thorough drying of the capacitor element before its soldering to the can, multilayer carbon coverage with high adhesion to the external silver layer to prevent delamination and silver migration toward the Ta_2O_5 dielectric, and high-temperature solder alloys to avoid whisker growth. To

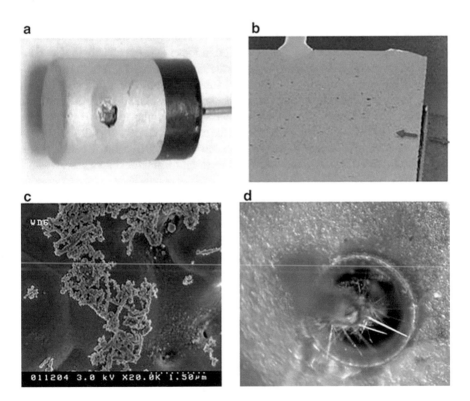

Fig. 3.49 Moisture-related failures in PHS Tantalum capacitors: popcorn effect (**a**), delamination of the external carbon and silver layers (**b**), silver migration (**c**), and tin whiskers (**d**)

improve humidity resistance of non-hermetic surface-mount Polymer Tantalum capacitors, especially in automotive applications, additional hydrophobic coatings in external layers of the capacitor elements and hydrophobic epoxy resin in the encapsulation are implemented.

In conclusion, the presence of moisture has a profound effect on asymmetric conduction and DCL stability in Polymer Tantalum capacitors. At normal polarity moisture acts as a plasticizer increasing polymer chain mobility and improving DCL stability due to maintaining a sufficient barrier at the Ta_2O_5/p-type PEDOT interface and activating self-healing processes in polymer cathode in vicinity of the structural defects in the Ta_2O_5 dielectric. At reverse polarity DCL stability degraded over time in presence of moisture due to H^+ ion diffusion toward the Ta/Ta_2O_5 interface, increasing current injection and eventually causing local overheating and unrepairable damage to the dielectric. The effects of moisture on Polymer Tantalum capacitors diminish with reducing overall size of tantalum anodes, usage of coarser tantalum powder and flawless dielectric technology (F-Tech) in the anode manufacturing, and reducing thickness of the Ta_2O_5 dielectric. These factors reduce the size and density of the structural defects in the Ta_2O_5 dielectric and improve coverage of the dielectric with polymer cathode.

Qualitatively similar effects of moisture on asymmetric conduction and stability are observed in Solid Electrolytic Tantalum capacitors with MnO_2 cathode. Particularly, at normal polarity moisture acts as a catalyst of the redox reaction at the MnO_2/Ta_2O_5 interface and phase transformations in the MnO_2 cathode in vicinity of the structural defects in the Ta_2O_5 dielectric, stabilizing DCL in these capacitors [10]. At reverse polarity, effect of moisture relates to H^+ ion diffusion toward the Ta/Ta_2O_5 interface provoking current instability in Solid Electrolytic Tantalum capacitors like that in Polymer Tantalum capacitors. Similar to Polymer Tantalum capacitors, reducing the size and density of the structural defects in the Ta_2O_5 dielectric and improving coverage of the Ta_2O_5 dielectric with MnO_2 cathode help increase BDV and improve DCL stability in Solid Electrolytic Tantalum capacitors.

3.5   Environmental Stability of AC Characteristics

Besides effects of moisture on BDV, DCL stability, asymmetric conduction, and moisture-related failures, it also affects AC characteristics of Polymer Tantalum capacitors, particularly, capacitance magnitude and its dependence on temperature and frequency. The effects of environmental conditions on capacitance stability with temperature, frequency, and bias voltage were investigated in both dry and humid PHS Tantalum capacitors with working voltages 15, 50, 75 and 100 V and corresponding oxide dielectric thicknesses (t_{ox}) 82.5, 275, and 412.5 nm [126, 127]. Figure 3.50 shows the capacitance dependence on temperature $C(T)$ in humid and dry PHS parts with different dielectric thicknesses (working voltages), measured at 120 Hz and zero bias voltage.

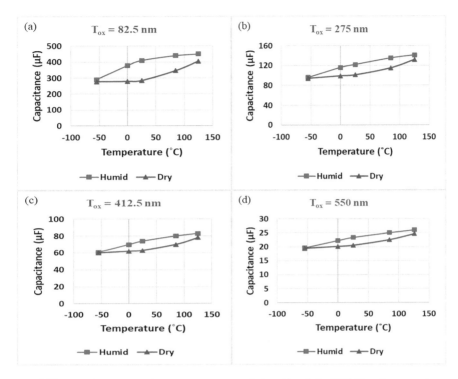

Fig. 3.50 Capacitance dependence on temperature in humid and dry PHS Tantalum capacitors measured at 120 Hz and 0 bias voltage for different dielectric thicknesses: (**a**) 82.5 nm, (**b**) 275 nm, (**c**) 412.5 nm, and (**d**) 550 nm

According to Fig. 3.50 for all dielectric thicknesses the capacitance in humid parts is higher than the capacitance in dry parts, and in all cases the capacitance increases with temperature. The difference in capacitance between humid and dry parts decreases at both low and high temperatures with respect to room temperature. In comparison to capacitance at room temperature, capacitance loss (decrease from room temperature value) at low temperatures is larger in humid parts than in dry parts while capacitance gain (increase from room temperature value) at high temperatures is higher in dry parts than in humid parts.

Figure 3.51 shows the relative change in capacitance with temperature with respect to capacitance at room temperature, $C(\mathrm{RT})$, which is calculated as $100 \times [C(T) - C(\mathrm{RT})]/C(\mathrm{RT})$ for humid and dry PHS parts with different dielectric film thicknesses.

According to Fig. 3.51, at low temperatures capacitance loss in humid parts is larger for parts with thinner dielectrics while capacitance loss in dry parts is low and approximately the same for all dielectric thicknesses. Conversely, at high temperatures capacitance gain in humid parts is low and about the same for all dielectric thicknesses while capacitance gain in dry parts is stronger for parts with thinner dielectrics.

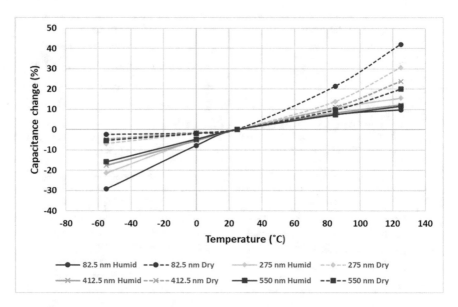

Fig. 3.51 Relative change in capacitance with respect to capacitance at room temperature in humid and dry PHS Tantalum capacitors

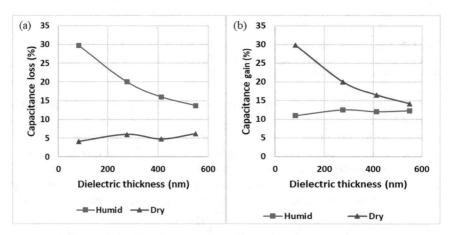

Fig. 3.52 Capacitance loss at −55 °C (**a**) and capacitance gain at 125 °C (**b**) with respect to capacitance at room temperature in humid and dry PHS Tantalum capacitors

The relative capacitance loss at $T = -55$ °C with respect to capacitance at room temperature is shown in Fig. 3.52a, which is calculated as $100 \times [C(RT) - C(-55$ °C)]$/C(RT)$. The capacitance gain at $T = +125$ °C with respect to capacitance at room temperature is shown in Fig. 3.52b, which is calculated as $100 \times [C(125$ °C$) - C(RT)]/C(RT)$. These capacitance loss and gain are plotted as a function of the dielectric thickness in both humid and dry PHS Tantalum capacitors.

According to Fig. 3.52, capacitance loss at −55 °C in dry parts and capacitance gain at 125 °C in humid parts are practically identical for different dielectric thicknesses. These losses and gains are similar to the temperature variation of the dielectric constant of the anodic Ta_2O_5 film [128]. However, capacitance losses at −55 °C in humid parts are much larger than in dry parts, while capacitance gains at 125 °C in dry parts are much larger than in humid parts, and in both cases the observed effect is more significant in devices with thinner dielectrics. Clearly the dielectric thickness plays a key role in these experimental results.

The results presented in Figs. 3.50, 3.51, and 3.52 show that there is a correlation between capacitance change from a dry condition C(dry) to a humid condition C(humid) and capacitance change with temperature in humid and dry parts. The relative capacitance change between humid and dry conditions as a function of temperature is shown in Fig. 3.53 and is calculated as $100 \times [C(\text{humid}) - C(\text{dry})]/C(\text{humid})$.

According to Fig. 3.53, the difference in capacitance between humid and dry conditions is maximum at room temperature and decreases at both low and high temperatures. Furthermore, this difference is larger for capacitors with the thinnest dielectric, decreasing for thicker dielectrics.

The effect of humidity on capacitance and its dependence on temperature in Polymer Tantalum capacitors can be related to the integrity of the dielectric-polymer interface [126]. When particles of the PEDOT slurry are deposited onto the dielectric surface, part of this surface does not have direct contact with the polymer (free area), and therefore does not contribute to the capacitance under dry conditions. Under humid conditions, the surface conductivity of the free area of the dielectric increases due to a thin layer of the absorbed molecules of water, which results in a higher total surface area and, thereby, a higher capacitance.

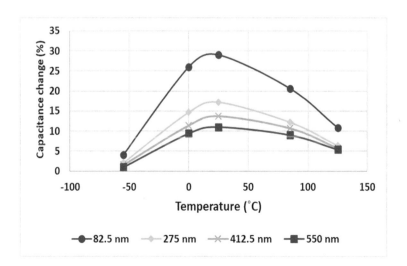

Fig. 3.53 Capacitance change between humid and dry PHS Tantalum capacitors as a function of temperature

At low temperatures, the surface conductivity of the dielectric in the free area of the humid parts decreases due to the low mobility of the ionic charge carriers, eventually resulting in a capacitance of humid parts that is practically equal to the capacitance in dry parts. However, the capacitance change with temperature in dry parts is low and correlates with the variation of the dielectric constant of the anodic Ta_2O_5 film. At high temperatures, the polymer in dry parts expands, resulting in an increase in surface area and, thereby, capacitance, which approaches the capacitance observed in humid parts. However, capacitance change with temperature in humid parts is low and correlates with the change in the dielectric constant of the anodic Ta_2O_5 film.

The stronger effect of environmental conditions on capacitance in capacitors with thinner dielectrics may be related to the larger surface roughness of the dielectric in these capacitors [127]. Figure 3.54 presents high-resolution SEM images of the surface of an unformed sintered tantalum anode and formed tantalum anodes with different anodic oxide film thicknesses.

From Fig. 3.54a, the unformed anodes, which have a 3.3 nm thick native oxide [49], demonstrate a relatively rough surface related to the crystalline structure on the tantalum particles in the sintered anode. As the thickness of the dielectric increases with increasing formation voltage, the dielectric surface becomes smoother as shown in Fig. 3.54b, c. This effect is due to the redistribution of the electric field in the anodic oxide film during anodization of tantalum [17]. For the same type of the tantalum anodes and polymer technology, coverage of the dielectric surface with the polymer improves as its surface becomes smoother, which naturally occurs for thicker films.

Capacitance dependence on frequency in Polymer Tantalum capacitors is also strongly affected by environmental conditions. Capacitance dependence on frequency in humid and dry PHS Tantalum capacitors at room temperature and zero bias voltage is shown in Fig. 3.55. Qualitatively similar results were obtained at other temperatures within the temperature range investigated in this work.

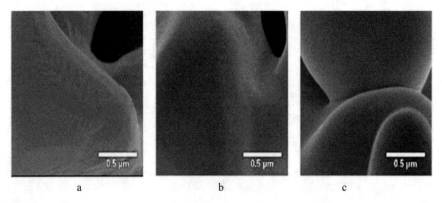

a   b   c

Fig. 3.54 SEM images of the unformed sintered tantalum anode (**a**), and tantalum anodes with 140 nm (**b**), and 292 nm (**c**) anodic oxide films

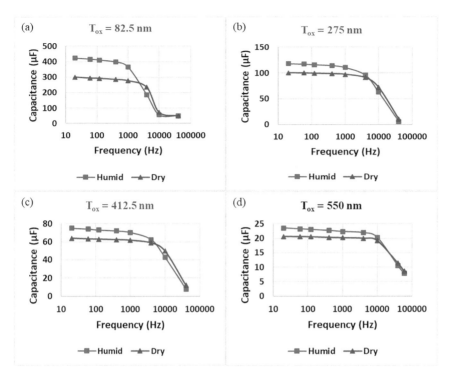

Fig. 3.55 Capacitance dependence on frequency in humid and dry PHS Tantalum capacitors for different dielectric thicknesses: (**a**) 82.5 nm, (**b**) 275 nm, (**c**) 412.5 nm, and (**d**) 550 nm

According to Fig. 3.55, at lower frequencies capacitance in humid parts reduces gradually with frequency while capacitance in dry parts remains relatively unchanged. Similar to capacitance loss at low temperatures in humid parts, capacitance loss with frequency in humid parts can be related to the low mobility of the ionic charge carriers which cannot follow the AC signal at higher frequencies. Eventually, at the "knee frequency" when capacitance in both humid and dry parts begins declining sharply, the capacitance of humid parts becomes equal to the capacitance of dry parts.

The "knee frequency" in tantalum capacitors with porous tantalum anodes has been explained by the J. Prymak's distributed capacitance model, which presents the capacitor as a sum of the capacitances of multiple layers starting with the external layer of the anode, followed by the dipper layer, and finally the core of the anode [101]. At the "knee frequency," the period of the AC signal, $T_p = 1/f$, becomes smaller than the time constant, $\tau = RC$, for the core of the anode, where R is the resistance of the core part of the anode, which is dominated by the resistance of the cathode. At $T_p < \tau$, the core part of the anode stops contributing to the total capacitance, and therefore the capacitance begins decreasing sharply. When the frequency continues to increase above the "knee frequency," the external layers of the anode, which have a lower resistance of the cathode compared to that of the core cathode due to the shorter distance to the anode surface, also stop contributing to the total capacitance which continues to decrease.

Similar to lower ESR, a higher "knee frequency" (higher capacitance stability with frequency) in Polymer Tantalum capacitors in comparison to Solid Electrolytic and Wet Tantalum capacitors can be explained by the higher conductivity (lower resistance) of the polymer cathode in comparison to the conductivity of the MnO_2 and liquid electrolyte cathodes. The difference in conductivity of these cathodes becomes larger at lower temperatures where there is a small change in conductivity of PEDOT, a p-type nearly degenerate semiconductor [129], some reduction in conductivity of MnO_2, an n-type semiconductor with a narrow band gap [130], and a strong reduction in conductivity of the liquid electrolyte approaching a gel state with minimal ionic mobility at −55 °C [131].

The relative change in capacitance with frequency is presented in Fig. 3.56, which is calculated as $100 \times [C(f) - C(20)]/C(20)$, where $C(20)$ is the reference capacitance at 20 Hz—the lowest frequency used in this work.

According to Fig. 3.56 the "knee frequency" is lower in Polymer Tantalum capacitors with thinner dielectrics and increases with increasing dielectric thickness. The ESR is also lower for the capacitors with thinner dielectric and increases with dielectric thickness; however, there is practically no difference in ESR between humid and dry PHS parts, as shown in Fig. 3.57.

The results presented in Figs. 3.56 and 3.57 show a lower "knee frequency" and lower ESR in Polymer Tantalum capacitors with thinner dielectrics. Furthermore, except for the 82.5 nm capacitors, there is only a small difference in knee frequency

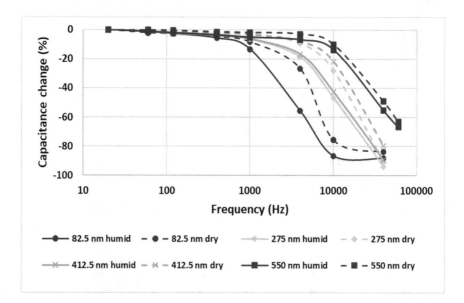

Fig. 3.56 Relative change in capacitance with respect to capacitance at 20 Hz in humid and dry PHS Tantalum capacitors

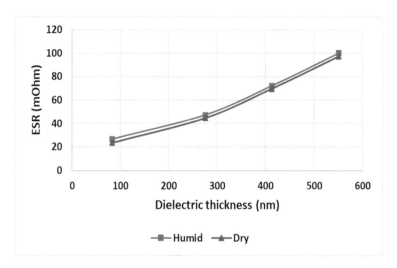

Fig. 3.57 ESR at 100 kHz for different dielectric thicknesses in humid and dry PHS Tantalum capacitors

between humid and dry PHS parts. There is also very little difference in ESR between humid and dry PHS parts for all dielectric thicknesses. These particular observations cannot be completely explained by the distributed capacitance model. For the same tantalum anode, Polymer Tantalum capacitors with thinner dielectrics have a higher capacitance and, thereby, a lower reactance $X_C = 1/2\pi fC$ as well as a lower self-resonance frequency $f = 1/2\pi(LC)^{1/2}$, where L is the parasitic inductance of the capacitor. The lower reactance and lower self-resonance frequency could explain the lower knee frequency in capacitors with thinner dielectrics in comparison to those with thicker dielectrics. Figure 3.56 shows that the frequency response curve for the 82.5 nm capacitors, especially for the humid samples, levels off at about 10 kHz and thereafter begins to increase. This behavior is expected at or near the self-resonance frequency. Furthermore, for all of the thicker dielectric samples, measurements showed that the curves level off and start to increase as the frequency continues to increase; however, these data were not plotted in Fig. 3.56 for the sake of clarity. A simple calculation estimates that the required parasitic inductance to reach self-resonance in these capacitors was a few tenths of a microhenry and only varied by about 0.5 μH across all thicknesses. Therefore, it seems reasonable that lower reactance and lower self-resonance frequency in the thinner dielectric Polymer Tantalum capacitors can adequately explain the results of Figs. 3.56 and 3.57.

The capacitance dependence on bias voltage in humid and dry PHS Tantalum capacitors at room temperature and 120 Hz is shown in Fig. 3.58. Qualitatively similar results were obtained at all other temperatures and frequencies.

According to Fig. 3.58, there is practically no capacitance dependence on bias for both humid and dry PHS Tantalum capacitors. Capacitance instability with bias voltage in dry Polymer Tantalum capacitors presented in [132] can be related to an increase in DCL at higher voltages. With high DC leakage in capacitors,

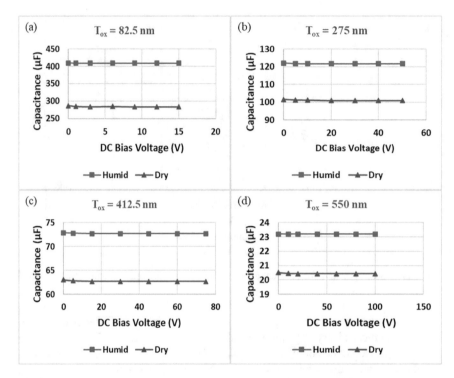

Fig. 3.58 Capacitance dependence on DC bias voltage in humid and dry PHS Tantalum capacitors for different dielectric thicknesses: (**a**) 82.5 nm, (**b**) 275 nm, (**c**) 412.5 nm, and (**d**) 550 nm

measurements of capacitance will be inaccurate since LCR meters are actually measuring displacement current and cannot distinguish between it and leakage current. Due to the flawless dielectric technology (F-Tech) and other advanced technologies and testing techniques used in the fabrication of PHS Tantalum capacitors, DCL was low and stable in both humid and dry parts at all voltages equal or less than the working voltage.

In conclusion, there is strong effect of environmental conditions on capacitance stability in Polymer Tantalum capacitors. Particularly, capacitance change with temperature $C(T)$ and frequency $C(f)$ correlates with capacitance change with humidification between dry and humid conditions. The coverage of the dielectric surface with slurry PEDOT particles plays a critical role in the effect of environmental conditions on these capacitors. Under humid conditions, the conductivity of the dielectric surface increases in the free area where there is no direct contact between the dielectric and polymer. This increase in surface conductivity of the dielectric, provided by the thin layer of absorbed water molecules, results in a higher total surface area and, thereby, a higher capacitance in humid parts in comparison to the capacitance in dry parts. At lower temperatures and higher frequencies, the conductivity of the humidified dielectric surface decreases due to the lower mobility of the ionic charge carriers, which increases capacitance loss with temperature and frequency in comparison to dry parts.

The effect of environmental conditions on capacitance stability in Polymer Tantalum capacitors is more pronounced in capacitors with thinner dielectrics as compared to capacitors with thicker dielectrics. This effect can be related to the changes in the dielectric surface with increasing thickness of the anodic Ta_2O_5 film, where the surface roughness decreases for thicker films but increases for thinner films. These changes in morphology of the dielectric surface affect the coverage of the dielectric with PEDOT particles and, thereby, the capacitance and its stability with temperature and frequency.

According to this model, improving the integrity of the dielectric-polymer interface (reducing the free area) should reduce the effect of environmental conditions and improve capacitance stability in Polymer Tantalum capacitors. There are several technological methods to achieve this goal [126]. One method is usage of a coarser tantalum powder in the tantalum anodes and, thereby, increasing the pore size for easier impregnation with polymer slurry particles. The disadvantage of this approach is reducing the volumetric efficiency (CV/cm^3), especially in lower voltage parts typically made with finer tantalum powder. Another method to improve the integrity of the dielectric-polymer interface is by using the in situ polymerization process instead of pre-polymerized PEDOT slurry. The molecules of PEDOT with in situ polymerization provide much better coverage of the dielectric surface in comparison to the slurry PEDOT particles. The frequency response also improves due to the smaller size molecules in in-situ PEDOT in comparison to the molecule size in slurry PEDOT. However, as it was shown earlier, in situ polymerization leaves residual by-products of the chemical reactions inside the polymer cathode, and these by-products affect DC characteristics, particularly BDV and DCL, especially in higher voltage parts with thicker dielectrics. Application of amine Silane to the dielectric surface prior to the slurry PEDOT also improves the dielectric coverage with slurry polymer particles without the negative effects on the volumetric efficiency, BDV, and DCL [133].

3.6 Anomalous Currents in Polymer Tantalum Capacitors

As it was presented earlier, lower-voltage tantalum capacitors with thinner dielectrics are typically manufactured with finer tantalum powder to increase the surface area of the anode and thereby the specific charge of the capacitor. In this situation, some pores in the sintered anodes become comparable in size or even smaller than particles of PEDOT in pre-polymerized PEDOT dispersions, which makes it practically impossible to impregnate these anodes with the pre-polymerized slurry PEDOT. That is why in situ PEDOT is still needed for internal impregnation of the porous anodes of lower-voltage Polymer Tantalum capacitors, while the pre-polymerized slurry PEDOT is applicable as an external part of the cathode in these capacitors. The hybrid polymer technology combines in situ internal PEDOT and pre-polymerized external PEDOT.

The hybrid PEDOT technology simplified and intensified the manufacturing process of lower-voltage Polymer Tantalum capacitors in comparison to pure in situ PEDOT technology; however, unexpected performance problems were revealed in these devices. Particularly, a very high transient current was detected when a short voltage pulse was applied after surface mounting of these capacitors on a circuit board [134]. As an example, Fig. 3.59 presents $I(t)$ and $V(t)$ characteristics for the first pulse and after repeated pulses at rated voltage and room temperature for a W-case 470 μF–6.3 V non-hermetic surface-mount Polymer Tantalum capacitor with anodes sintered with 150,000 μC/g tantalum powder, Ta_2O_5 dielectric formed at 17 V, and hybrid PEDOT technology.

As one can see from Fig. 3.59a, current begins increasing sharply after a certain delay when voltage is applied. This anomalous transient current reaches a maximum value in the ampere range and then decays to the milliamp range within a fraction of a second. The smaller spikes due to pure displacement current can be seen at the beginning of the pulse. The magnitude of transient current decreases after each pulse and ultimately becomes negligible after repeated testing, as seen in Fig. 3.59b. The behavior demonstrated by the hybrid Polymer Tantalum capacitors suggests that when the field is applied to the capacitor, PEDOT-PSS macromolecules carrying strong dipoles reorient accordingly. This reorientation occurs with a certain rate connected to the mobility of the polymer segments. If the reorientation rate is significantly lower than the rate of the charge application, an anomalously high current is observed during the first pulse. Since application of the following pulses does not result in significant current, it can be concluded that the chains are settled in their new quasi-permanent positions after the very first pulse.

The response rate of the polymer chains can be regulated by control of the polymer chain mobility [135]. First of all, the mobility can be increased by the addition of small molecules, plasticizers, that dissolve in the polymer, separating the chains from each other and hence making the chain movement easier. It was determined that the polymer cathode absorbed an amount of water equal to about 8% of the dry polymer weight when exposed to the air with a relative humidity of about 70% for several hours. In fact, no measurable anomalous current was detected in hybrid Polymer Tantalum capacitors after the plasticizing water molecules were introduced into the system. Drying of the capacitors at 125 °C for 24 h restored the anomalous transient current similarly to that shown in Fig. 3.52a. DCL in these capacitors measured after rated voltage was applied for 90 s at room temperature was low and practically identical before and after voltage pulses were applied.

Secondly, the chain mobility can be significantly decreased by decreasing the temperature of the sample. Figure 3.60 shows the effect of temperature on anomalous transient current in W-case 470 μF–6.3 V hybrid Polymer Tantalum capacitor.

According to Fig. 3.60, the time delay at the beginning of the transient current increase, the magnitude of the peak current, and the amount of charge corresponding to the transient current (area under the current curve) all increase with decreasing temperature. The highest magnitude of the transient current as well as the

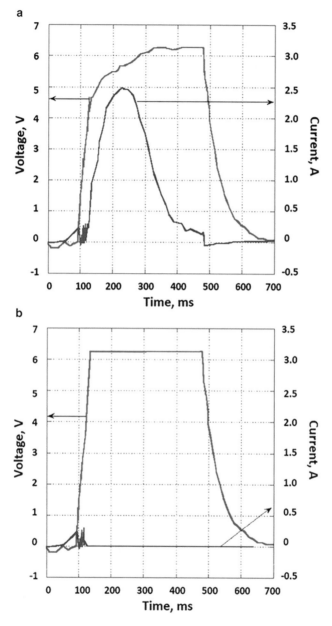

Fig. 3.59 $I(t)$ response to first $V(t)$ pulse (**a**) and after repeated pulses (**b**) applied at room temperature to a W-case 470 μF–6.3 V hybrid Polymer Tantalum capacitor

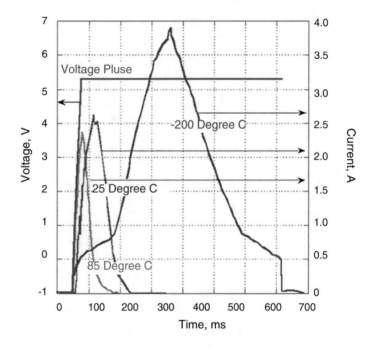

Fig. 3.60 Effect of temperature on anomalous transient current, $I(t)$, in response to a voltage pulse, $V(t)$, applied to a W-case 470 μF–6.3 V hybrid Polymer Tantalum capacitor

amount of the corresponding charge were detected at −200 °C, after the capacitor was cooled by liquid nitrogen.

The magnitude of the transient current can be also increased by applying voltage pulse at reverse polarity (minus on tantalum anode). The first pulse at normal polarity following the pulse at reverse polarity shows high transient current similar to that shown in Fig. 3.59a even though the magnitude of the transient current was negligible prior to the reverse pulse similar to that shown in Fig. 3.59b. From these data, we observe that high anomalous transient current in Polymer Tantalum capacitors, which diminishes after repetitions of pulses at normal polarity, can be restored by a short application of reverse voltage without long-term drying at high temperatures. It is evident that application of a reverse bias causes the migration of the polymer segments to their initial (prior to the first pulse) positions as indicated by the significant current observed.

The anomalous transient currents were observed in hybrid Polymer Tantalum capacitors, while practically no anomalous transient currents were observed in pure in situ Polymer Tantalum capacitors. Figure 3.61 shows the pulse characteristics of W-case 470 μF–6.3 V Polymer Tantalum capacitors with hybrid and pure in situ cathodes, obtained after drying the capacitors at 125 °C for 24 h and then cooling them to −200 °C in liquid nitrogen.

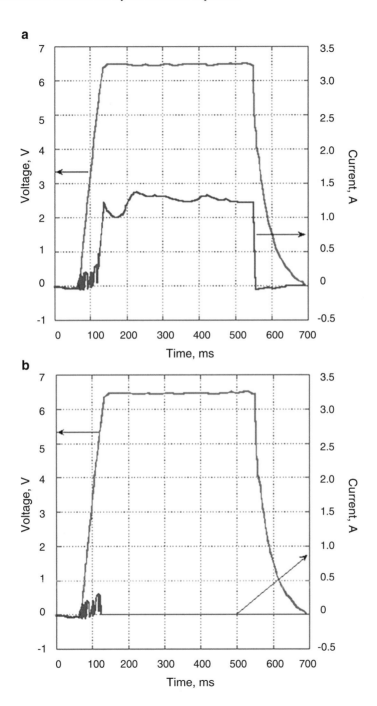

Fig. 3.61 $I(t)$ response to one pulse, $V(t)$, applied at $-200\ °C$ to a W-case 470 µF–6.3 V hybrid (**a**) and pure in situ (**b**) Polymer Tantalum capacitors

According to Fig. 3.61, there is anomalous transient current at rated voltage only in hybrid Polymer Tantalum capacitors with pre-polymerized PEDOT-PSS molecules in external slurry polymer cathode (Fig. 3.61a), while there is practically no anomalous transient current in Polymer Tantalum capacitors with pure in situ PEDOT cathode (Fig. 3.61b).

The DCL characteristics are also different in Polymer Tantalum capacitors with hybrid and pure in situ PEDOT cathodes [134]. Figure 3.62 presents I–V characteristics of W-case 470 μF–6.3 V hybrid and pure in situ Polymer Tantalum capacitors at room temperature, 85 °C, and liquid nitrogen temperature.

According to Fig. 3.62a, in hybrid capacitors at normal polarity anomalously high DCL was detected at −200 °C. The difference between DCL at −200 °C and DCL at room temperature and 85 °C is increasing with voltage. The high current in the range of several milliamps was observed in the hybrid capacitors for several hours as long as the capacitors remained submerged in liquid nitrogen and rated voltage at normal polarity was applied. However, in capacitors with pure in situ cathodes (Fig. 3.62b), DCL at normal polarity is lower at −200 °C in comparison to DCL at room temperature and 85 °C. Both hybrid PEDOT and in situ PEDOT types

Fig. 3.62 I–V curves of W-case 470 μF–6.3 V hybrid (**a**) and in situ (**b**) Polymer Tantalum capacitors at −200, 25, and 85 °C

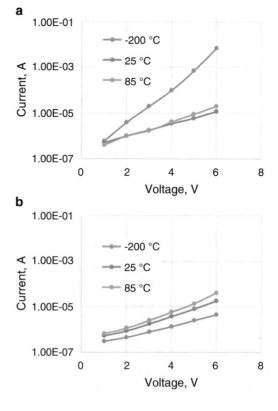

of capacitors demonstrate polar behavior with low current at normal polarity in comparison to the much higher current at reverse polarity.

The difference in the behavior between the hybrid and in situ Polymer Tantalum capacitors can be connected to the discussed earlier possibility for the polymer chains, constituting the pre-polymerized slurry PEDOT-PSS cathode, to reorient more slowly when the electrical field is applied. The reorientation influences the band structure at the dielectric-polymer interface, where the mobile dipoles orient themselves according to the field. When the PEDOT-PSS cathode is humidified, the reorientation happens more quickly, since the water works as a plasticizer for the polymer chains involved. When the PEDOT-PSS cathode is dry, the movement/ reorientation of the dipoles becomes slower, which results in the significant anomalous current. A decrease in temperature decreases the mobility of the dipoles as well. For the in situ capacitors, there are no polar PSS macromolecules present. In this case only low molecular weight paratoluene sulfonic acid (pTSA) is associated with PEDOT units, and the water molecules have a much smaller impact on the transient current of in situ Polymer Tantalum capacitors vs. hybrid ones.

The difference in polymer cathodes can also explain the difference in the behavior between the hybrid and in situ Polymer Tantalum capacitors during the breakdown voltage (BDV) testing. As an example, Fig. 3.63 shows $I–V$ and $C–V$ curves during BDV testing of the W-case 470 μF–6.3 V Polymer Tantalum capacitors with either hybrid or pure in situ cathode. The current and capacitance were registered at room temperature, while voltage increased at a constant rate of 0.5 V/min from 0 V to the voltage at which the 1 A fuse opened or to a maximum value of 25 V, which is about 50% higher than formation voltage 17 V.

The results presented in Fig. 3.63 show that breakdown for in situ capacitors (Fig. 3.63b) was characterized by a rapid current increase and capacitance loss as the applied voltage approached formation voltage. Furthermore, these capacitors effectively became shorts and the fuse was blown. For hybrid capacitors however, a relatively low current and gradual decrease in capacitance were observed as applied voltage achieved formation voltage and continued increasing (Fig. 3.63a). The different types of breakdown observed for in situ and hybrid Polymer Tantalum capacitors can be also explained by the differences in their polymer cathodes. At extreme voltages, approaching BDV, PSS chains in the hybrid capacitors can separate from the PEDOT chains and form an additional dielectric layer on top of the tantalum oxide dielectric. This results in a loss of capacitance and relatively low-leakage current even when the applied voltage exceeds the formation voltage [136]. Low conductivity of undoped PEDOT separated from PSS also helps decrease leakage current at extreme voltages.

Quality of the dielectric also plays important role in anomalous currents in higher-voltage Polymer Tantalum capacitors with hybrid or pure slurry PEDOT cathode, particularly in the anomalous charge current when rapid voltage ramp is applied to the capacitor [105]. As an example, Figure 3.64 shows the charge current distributions at voltage ramp test in D-case 15 μF–35 V Polymer Tantalum capacitors manufactured with pure slurry PEDOT cathodes and either conventional dielectric technology or flawless dielectric technology (F-Tech). The parts were baked out

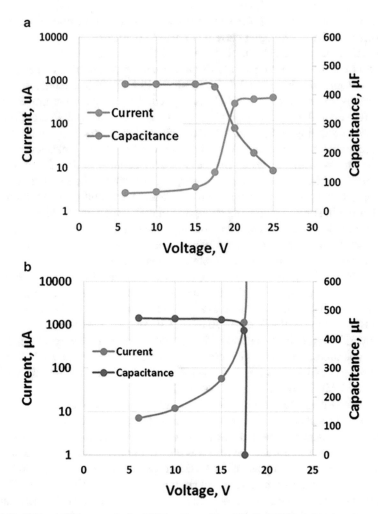

Fig. 3.63 *I–V* and *C–V* curves during BDV testing of the 470 µF–6.3 V hybrid (**a**) and pure in situ (**b**) Polymer Tantalum capacitors

in the air at 125 °C for 16 h, cooled down to 0 °C, and charged immediately to 28 V with voltage ramp $dV/dt = 120$ V/s.

As one can see in Fig. 3.64, the average charge current in Polymer Tantalum capacitors manufactured with conventional dielectric technology (Fig. 3.64a) is about 10× the theoretical charge current $I_{th} = 1.8$ mA calculated as $I_{th} = C \times dV/dt$, while it is about 2.5× the theoretical current in the Polymer Tantalum capacitors manufactured with F-Tech (Fig. 3.64b).

In conclusion, the anomalous currents at normal polarity (+ on the tantalum anode) including anomalous transient current when a pulse of rated voltage is applied to the capacitor, anomalous charge current when rapid voltage ramp is

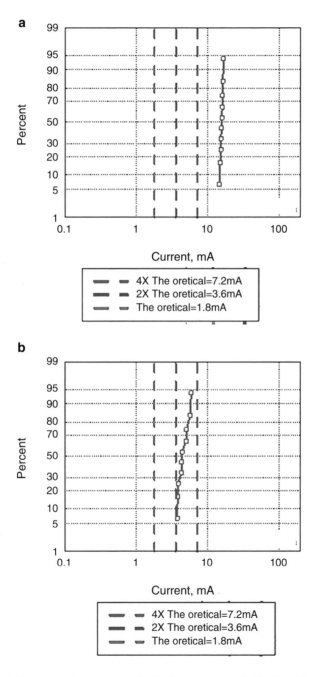

Fig. 3.64 Probability plots for the current distributions in D-case 15 μF–35 V Polymer Tantalum capacitors with slurry PEDOT cathodes and either conventional dielectric technology (**a**) or F-Tech (**b**) charged to 28 V with a ramp of 120 V/s

applied to the capacitor, and also anomalous DC current when rated voltage is applied to the capacitors at low temperature, were detected in Polymer Tantalum capacitors with hybrid in situ—slurry or pure slurry PEDOT cathode. Anomalous currents decrease with the humidification and pulse repetitions, but they can be restored by drying the capacitors or by application of the voltage pulse at reverse polarity (− on tantalum anode).

The anomalous currents observed in Polymer Tantalum capacitors can be explained by the presence of dipoles, charged polymer chains, in the conducting polymer cathode at its interface with the oxide dielectric. These dipoles will reorient when an electrical field is applied, and this reorientation determines the potential barrier at the oxide-polymer interface, which plays a crucial role in limiting the current through the Polymer Tantalum capacitor. The dipole orientation time is a function of temperature, and therefore the process slows down at lower temperatures when mobility of polymer chains becomes low. In humidified capacitors the reorientation is faster than the process in dry capacitors, since the water acts as a plasticizer for the polymer chains. This reorientation, and the resulting anomalous currents, is more pronounced in pre-polymerized slurry PEDOT cathodes with long molecules of PSS in the polymer structure; therefore, it is more pronounced in capacitors with the hybrid or pure slurry polymer cathodes as compared to those with pure in situ polymer cathodes. The differences in the polymer structure in the lower-voltage capacitors with in situ and hybrid polymer cathodes can also explain the differences in their electric breakdown and self-healing properties.

In the higher-voltage Polymer Tantalum capacitors with hybrid and pure slurry PEDOT cathode, quality of the tantalum oxide dielectric also plays important role in the anomalous currents at normal polarity (plus on tantalum anode). Particularly flawless dielectric technology (F-Tech) provides significant reduction in anomalous charge current when rapid voltage ramp is applied to the capacitor after thorough drying in comparison to that in Polymer Tantalum capacitors manufactured with conventional dielectric technology. Continuous improvements in the structure and chemical composition of the tantalum oxide dielectric and polymer cathode will further reduce the anomalous currents in Polymer Tantalum capacitors with hybrid and pure slurry PEDOT cathode.

3.7 High-Temperature Applications

Increasing operating temperatures in electronic components to 175–200 °C and higher is desirable for applications such as downhole oil and gas exploration, geothermal energy generation, power electronics, under the hood electronics, military devices, and avionics [94, 137, 138]. These applications also involve strong vibration and shocks, which require mechanical robustness in addition to ability to operate at high temperatures. To withstand these harsh application conditions, all the materials involved in the electronic components and the interfaces between different materials should have sufficient thermal and mechanical strength. For example, in traditional ceramic capacitors, the dielectric is based on barium neodymium titanate and compatible with precious metal electrode such as palladium or silver-palladium alloy, while in high-temperature ceramic capacitors, the dielectric was converted to calcium zirconate material compatible with base metal, mostly nickel, electrodes [138].

In comparison to ceramic capacitors, no changes in materials comprising basic structure of tantalum capacitors, Ta-Ta_2O_5-cathode, can be made to adjust to high temperatures. Moreover, high-temperature applications rule out Wet Tantalum capacitors with liquid electrolyte cathodes due to excessive vapor pressure inside hermetic cans as well as Polymer Tantalum capacitors due to glass transition and decomposition of the conductive polymer cathode. Only Solid Electrolytic Tantalum capacitors with manganese dioxide cathodes potentially can withstand high-temperature application. Manufacturing of Solid Electrolytic Tantalum capacitors includes numerous heat treatments at temperatures equal or exceeding 260 °C during pyrolytic deposition of MnO_2 cathode on the surface of the Ta_2O_5 dielectric. Thermal gravitometer analysis (TGA) of pyrolytic MnO_2 showed stable weight up to 580 °C while losing weight, decomposition, at higher temperatures [139]. When separated from tantalum substrate, Ta_2O_5 dielectric shows no change in its chemical composition and structure until about 600 °C with phase transformation from amorphous to crystalline structure at higher temperatures [19]. With thermally stable individual layers in basic structure Ta-Ta_2O_5-MnO_2, stable behavior of the Solid Electrolytic Tantalum capacitors can be expected at temperatures 300 °C or even higher temperatures; however, experimental data contradict this assumption.

Capacitance and ESR stability with time for 1000 h testing at 260 °C without voltage applied were investigated in Solid Electrolytic Tantalum capacitors from different manufacturers [140]. Despite some quantitative differences, all the capacitors demonstrated a short initial period of the capacitance increase and relatively stable ESR followed by continues capacitance loss and ESR increase with time over the duration of the test. These experimental data can be explained by kinetics of the oxygen migration in the basic structure Ta-Ta_2O_5-MnO_2. The oxygen migration is initiated at the Ta/Ta_2O_5 interface where tantalum anode spontaneously extracts oxygen from the dielectric, enriching tantalum with oxygen and leaving oxygen vacancies in the dielectric. As it was discussed in Sect. 1.1, the oxygen vacancies increase conductivity of the Ta_2O_5 dielectric near the Ta-Ta_2O_5 interface, reducing effective thickness of the dielectric and, thereby, increasing capacitance at the initial stage of the high-temperature testing. Under the gradient of concentration, oxygen vacancies diffuse toward the Ta_2O_5-MnO_2 interface where they are compensated by oxygen from the MnO_2 cathode. Gradient concentration of oxygen causes oxygen diffusion in the MnO_2 cathode toward its interface with the dielectric. Eventually oxygen migration in basic structure Ta-Ta_2O_5-MnO_2 includes the chain of diffusion and redox processes such as oxygen diffusion in MnO_2 cathode, redox reaction at the Ta_2O_5-MnO_2 interface, oxygen diffusion in Ta_2O_5 dielectric, redox reaction at the Ta-Ta_2O_5 interface, and oxygen diffusion in tantalum anode. The relative intensity of the individual processes in this chain depends on temperature and determines long-term stability of the Solid Electrolytic Tantalum capacitors [19].

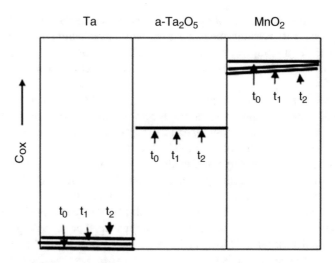

Fig. 3.65 The qualitative scheme of the time-dependent ($t_0 > t_1 > t_2$) variation in oxygen concentration (C_{ox}) in Ta-Ta$_2$O$_5$-MnO$_2$ structure at normal operating temperatures

Figure 3.65 shows the qualitative scheme of the time-dependent variation in oxygen concentration in Ta-Ta$_2$O$_5$-MnO$_2$ structure with amorphous dielectric (a-Ta$_2$O$_5$) at normal operating temperatures.

In this case, redox reaction at the Ta-Ta$_2$O$_5$ interface, initiating the chain of the oxygen migration processes, is slower than the oxygen diffusion in tantalum and in the Ta$_2$O$_5$ dielectric. Therefore, the concentrations of oxygen on both sides of this interface are represented as practically uniform. The oxidation strength of the MnO$_2$ cathode balances fully the oxygen deficiency in the Ta$_2$O$_5$ dielectric resulting from the extraction of oxygen by tantalum. The concentration of oxygen in the MnO$_2$ cathode is almost homogeneous, and the influx of oxygen in the dielectric is achieved due to the whole MnO$_2$ thickness. Here lies the reason for the long-term dynamic stability of the chemical composition of the Ta$_2$O$_5$ dielectric and, thereby, stable electrical parameters of the Solid Electrolytic Tantalum capacitors at normal operating temperatures.

Principally different time-dependent variation in oxygen concentration in Ta-Ta$_2$O$_5$-MnO$_2$ structure takes place at high operating temperatures (Fig. 3.66).

In this case, the oxygen extraction from the Ta$_2$O$_5$ dielectric into the tantalum anode is so active that it can't be balanced by the influx of oxygen into the dielectric from the MnO$_2$ cathode. As a result, oxygen deficit x in the Ta$_2$O$_{5-x}$ dielectric is increasing with time. Eventually phase transformations take place in tantalum anode, where oxygen concentration reaches the solubility limit of oxygen in tantalum, Ta(O), and crystalline Ta$_2$O$_5$ (cr-Ta$_2$O$_5$) phase precipitates at the anode-dielectric interface. The phase transformations also take place in MnO$_2$ cathode, where a primary oxygen depletion of the cathode layer adjacent to the dielectric causes precipitation of the lower manganese oxide phases such as Mn$_2$O$_3$. Figure 3.67 shows the qualitative scheme of the phase composition of the basic capacitor

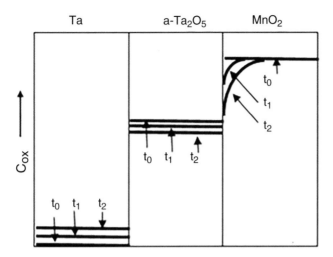

Fig. 3.66 The qualitative scheme of the time-dependent variation in oxygen concentration in Ta-Ta_2O_5-MnO_2 structure at high operating temperatures

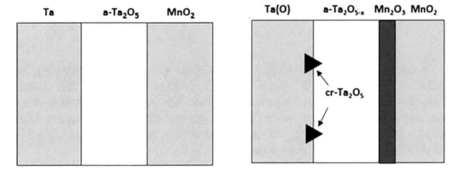

Fig. 3.67 The qualitative scheme of the phase composition of the basic capacitor structure before and after the long-term testing of the Solid Electrolytic Tantalum capacitor at high operating temperature

structure before and after the long-term testing of the Solid Electrolytic Tantalum capacitor at high operating temperatures.

As shown in Fig. 3.67b, phase transformations in basic structure of the Solid Electrolytic Tantalum capacitors in vicinity of the anode-dielectric and dielectric-cathode interfaces were discovered experimentally by electron diffraction analysis of the heat-treated capacitor structures built on the sputtered film of tantalum as anode [19]. The development of the low manganese oxide with high resistivity at the dielectric-cathode interface increases the effective thickness of the dielectric, resulting in monotonic capacitance loss and ESR increase during the long-term testing of the Solid Electrolytic Tantalum capacitors at high temperatures as it was shown in [140]. The depletion of the Ta_2O_5 dielectric with oxygen and growth of the crystalline Ta_2O_5 inclusions at the anode-dielectric interface cause DC leakage increase and eventually result in parametric and catastrophic failures of Solid Electrolytic Tantalum capacitors.

The experimental results show that the transition from normal to high operating temperatures scenarios in Solid Electrolytic Tantalum capacitors occurs between 200 °C, allowing long-term reliable application with temperature-related de-rating, and 260 °C, when the electrical parameters degrade quickly with time even without voltage applied. According to [19] there is no measurable oxygen flow in Ta-Ta_2O_5-MnO_2 structure at 200 °C, while there is significant oxygen flow at 260 °C and higher temperatures.

The oxygen current I_o at temperature T can be estimated using Arrhenius equation:

$$I_o = I_{oc}{}^* \exp-\left(E_a / kT\right),$$

where I_{oc} is the material-related constant, E_a the activation energy of the process, and $k = 8.6E{-}05$ eV/K the Boltzmann constant. From this equation, the ratio between oxygen flow I_1 at temperature T_1 and I_2 at temperature T_2 can be presented as:

$$I_1 / I_2 = \exp\left[E_a\left(T_2 - T_1\right) / kT_1T_2\right]$$

Assuming $T_1 = 473$ K (200 °C), $T_2 = 533$ K (260 °C), and $E_a = 1.82$ eV [19], the ratio I_1/I_2 calculated from this equation is approximately $I_1/I_2 = 1.7e^{18}$. This estimation shows extremely rapid increase in oxygen flow as operating temperatures increase above 200 °C. Applying voltage at normal polarity accelerates migration of oxygen ions and charged oxygen vacancies; however, effect of applied voltage is diminished by significant temperature-related de-rating.

Described earlier technologies that suppress degradation processes in tantalum capacitors, such as oxygen migration and crystallization of the amorphous matrix of the Ta_2O_5 dielectric, help improve thermal stability of the basic structure of the Solid Electrolytic Tantalum capacitors. These technologies include F-Tech and relatively coarse tantalum powder in the anode manufacturing, doping of the Ta_2O_5 dielectric with phosphorus during formation process, full coverage of the Ta_2O_5 dielectric with MnO_2 cathode inside and outside porous anodes, etc. Sufficiently thick and dense cathode layer on the surface of the Ta_2O_5 dielectric improves self-healing properties of the Solid Electrolytic Tantalum capacitors and, thereby, stabilizes basic structure of these capacitors at high operating temperatures.

Besides the basic structure Ta-Ta_2O_5-MnO_2, the external layers of carbon and metallization, applied to the surface of the MnO_2 cathode, also play important role in increasing the operating temperatures of the Solid Electrolytic Tantalum capacitors. Temperature robust organic binders in carbon and silver formulations provide sufficient stability to the external carbon and silver layers at temperatures well exceeding 200 °C [139]. Nevertheless, at high operating temperatures, silver can migrate through the micropores in carbon layer and external MnO_2 cathode and deposit on the surface of the Ta_2O_5 dielectric, causing high DCL and the capacitor failure [116]. The presence of silver on the surface of the Ta_2O_5 dielectric in Solid Electrolytic Tantalum capacitors exposed to long-term storage at 230 °C was detected experimentally using scanning electron microscopy (SEM) energy dispersive spectrometry (EDS) [138].

According to [116], local metal-oxide-metal (MOM) structures Ta-Ta_2O_5-Ag cause DCL increase when silver deposits on the defect areas in the dielectric. According to [110] breakdown voltage in the Ta-Ta_2O_5-Ag structures is relatively low even in the defect-free areas of the dielectric and practically constant as the formation voltage exceeds about 50 V. Low BDV in the Ta-Ta_2O_5-Ag structures was explained in [116] by the fact that Ta_2O_5-Ag interface does not form potential barrier for the current carriers and does not provide self-healing to the dielectric. That's why deposition of even small amount of silver on the surface of the Ta_2O_5 dielectric can cause catastrophic failures in the Solid Electrolytic Tantalum capacitors. In presence of humidity, silver migration is dominated by transport of positively charged silver ions, which is active during storage and, especially, reverse voltage testing [116]. At the same time, at high temperatures silver migration can be dominated by diffusion of neutral silver atoms, which is not affected by the presence of humidity or polarity of applied voltage.

As an alternative metallization, silver can be replaced by nickel, plated on the top of the external carbon layer by applying reverse voltage to the tantalum anode and running reverse current through the dielectric of the capacitor element partially submerged in aqueous solution of a nickel salt [141]. The external nickel layer improves thermal stability of the Solid Electrolytic Tantalum capacitors due to lower diffusion activity of nickel in comparison to silver, which prevents metal migration from the external metallization to the surface of the Ta_2O_5 dielectric. This technology is limited to lower-voltage tantalum capacitors, since applying reverse voltage can cause permanent damage to the Ta_2O_5 dielectric, especially, in higher-voltage capacitors with thicker dielectrics (Sect. 3.4). Sprague Electric, the original manufacturer of the Solid Electrolytic Tantalum capacitors, used sprayed copper on top of the carbon layer as the external metallization [5]. Similar to plated nickel, sprayed copper has low diffusion activity in comparison to silver and doesn't require an organic binder. At the same time, spraying copper doesn't affect the quality of the Ta_2O_5 dielectric regardless of dielectric thickness. To improve the uniformity of the external copper layer and make the process more manufacturing friendly, copper spraying can be replaced with physical vapor deposition (PVD) or magnetron sputtering [142].

Besides replacing silver with nickel or copper in the external metallization, the silver adhesive used to attach the capacitor element to the lead frame can be also replaced with transient liquid phase sinterable material. This material has relatively low melting temperature but once cured forms alloy with much higher melting temperature well exceeding the highest possible operating temperatures of the Solid Electrolytic Tantalum capacitors [143, 144]. The molding compound used to encapsulate the surface-mount tantalum capacitors as well as external terminations can be also optimized for high-temperature applications [139]. As an example, Fig. 3.68

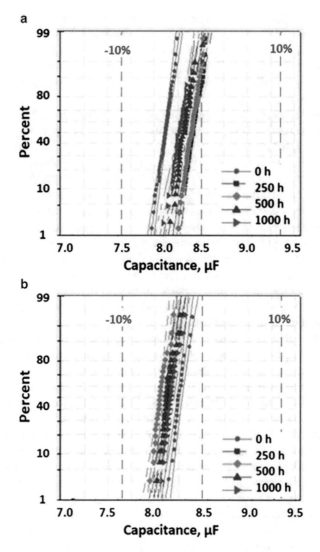

Fig. 3.68 Capacitance distribution during the 1000 h storage at 250 °C (**a**) and life test at 250 °C and 10 V (**b**) for molded D-case 8.5 μF–45 V Solid Electrolytic Tantalum capacitors

presents capacitance distribution during the 1000 h storage at 250 °C and life test at 250 °C and 10 V for molded D-case 8.5 μF–45 V Solid Electrolytic Tantalum capacitors manufactured with these technologies and materials.

As one can see in Fig. 3.68, capacitance variations during the long-term storage and life test at 250 °C were within +/− 10%, which indicates no phase transformation in the basic structure Ta-Ta_2O_5-MnO_2 and shows a possibility for stable application of the Solid Electrolytic Tantalum capacitors even at such a high operating temperature.

In conclusion, Solid Electrolytic Tantalum capacitors allow higher operating temperatures in comparison to Wet and Polymer Tantalum capacitors. Oxygen migration in basic structure Ta-Ta_2O_5-MnO_2 and related phase transformations in individual layers of the basic structure near the anode-dielectric and dielectric-cathode interfaces play critical role in stability of the Solid Electrolytic Tantalum capacitors at high operating temperatures. These processes activate rapidly as operating temperatures exceed 200 °C. Silver migration from the external silver metallization layer and silver adhesive toward the Ta_2O_5 dielectric also causes degradation and failures of the Solid Electrolytic Tantalum capacitors at high operating temperatures. High-temperature-related de-rating requires usage of the high-voltage tantalum capacitors even for the low-voltage applications, while thick Ta_2O_5 dielectrics in the high-voltage tantalum capacitors are especially susceptible to the degradation and failure at high operating temperatures. Only special technologies that suppress oxygen migration and phase transformations in basic structure Ta-Ta_2O_5-MnO_2 and materials that eliminate metal migration toward the surface of the Ta_2O_5 dielectric and provide thermally stable encapsulation and corrosion-resistant external terminations allow manufacturing of the Solid Electrolytic Tantalum capacitors for reliable applications at temperatures well exceeding 200 °C.

3.8   Niobium and Niobium Oxide Capacitors

As it was presented earlier, development of first niobium capacitors in former USSR was motivated by shortage of tantalum there, while niobium was plentifully available. The Cold War limited USSR the ability to use external sources of tantalum. The OJSC ELECOND in Russia continues mass manufacturing of Solid Electrolytic niobium capacitors with niobium (Nb) anode, niobium pentoxide (Nb_2O_5) dielectric and MnO_2 cathode for commercial and special applications with working voltages 6.3–25 V and capacitance 0.22–680 μF. The recent interest in niobium-based capacitors was driven by worldwide shortage of tantalum in early 2000s due to the vast consumption of the tantalum capacitors in the fast-growing telecommunication industry. In 2003, the AVX Corporation launched commercial niobium oxide capacitors with niobium monoxide (NbO) anode, Nb_2O_5 dielectric and MnO_2 cathode, and working voltages 1.5–6.3 V (later 10 V) and capacitance 1–1000 μF [21]. Comparison of the physical and electrical parameters of the anode and dielectric in tantalum, niobium, and niobium oxide capacitors is presented in Table 3.1.

In this table d is the density of the anode material, R is the anode resistivity, k is dielectric constant, and a is formation constant defined as $t_{ox} = aV_f$, where t_{ox} is the dielectric thickness and V_f is formation voltage (all at 20 °C).

As shown in the table, higher resistivity of NbO in comparison to the resistivity of tantalum and niobium causes some increase of ESR in niobium oxide capacitors in comparison to ESR in the same size tantalum and niobium capacitors. At the same time, resistivity of the MnO_2 cathode, which is about 0.1–1 Ohm/cm, is

Table 3.1 Physical and electrical parameters of anode and dielectric in tantalum, niobium, and niobium oxide capacitors

Type	Anode			Dielectric		
	Symbol	d (g/cm^3)	R (Ohm/cm)	Symbol	k	a (nm/V)
Tantalum	Ta	16.6	$(1–12) \times 10^{-6}$	Ta_2O_5	27	1.6
Niobium	Nb	8.57	$(4–15) \times 10^{-6}$	Nb_2O_5	41	2.4
Niobium Oxide	NbO	7.26	$(1.3–2) \times 10^{-2}$	Nb_2O_5	41	2.4

significantly higher than resistivity of the NbO anode and dominates ESR in niobium oxide capacitors.

The capacitance C per unit of anode surface area A is directly proportional to the dielectric constant k and inversely proportional to the volt constant a:

$$C / A = n^* k / a,$$

where $n = k_o/V_f$ and k_o is the permittivity of vacuum. From the k and a data presented in the table, the ratio k/a is approximately 1.7 V/nm in tantalum, niobium, and niobium oxide capacitors. In this case, at given formation voltage, all three types of the capacitors have approximately the same value of the capacitance per unit of anode surface, or, in other words, increase in the dielectric constant in the Nb_2O_5 dielectric in comparison to the Ta_2O_5 dielectric is offset by the larger thickness of the Nb_2O_5 dielectric in comparison to the thickness of the Ta_2O_5 dielectric.

The volumetric efficiency CV/g is inversely proportional to the anode density, which is typically about 1/3 of the density of the solid anode material. From the table, density of tantalum is approximately 1.9× density of niobium and 2.3× density of NbO. In this case, volumetric efficiency of the NbO anodes sintered with 100,000 CV/g NbO powder [21] can be compared with volumetric efficiency of tantalum anodes sintered with 43,500 CV/g tantalum powder. At the same time, low-voltage tantalum capacitors are currently manufactured with anodes sintered with 100,000–250,000 CV/g tantalum powders, which provides two to five times higher volumetric efficiency to tantalum anodes in comparison to Nb and NbO anodes. Since the anodes occupy the main part of the volume of the finished capacitors and the ratio between the formation voltage and rated voltage is approximately the same in all three types of the capacitors, low-voltage tantalum capacitors have approximately two to five times smaller volume in comparison to the niobium and niobium oxide capacitors with the same capacitance and voltage.

The higher ignition energy of the NbO anodes with 50% atm of oxygen in comparison to ignition energy of tantalum and niobium anodes with much lower oxygen content was presented as the safety feature of the niobium oxide capacitors [21]. At the same time according to Sect. 3.2, there is no ignition of tantalum anodes in failed tantalum capacitors as well. The actual safety feature of the niobium oxide capacitors is relatively high resistance after failure [21]. This resistance depends on the CV of the niobium oxide capacitor and maximum current in the electric circuit [22]. As an example, Fig. 3.69 shows resistance of the D-case 150 μF–10 V niobium oxide capacitors after BDV test with either 1 or 5 A fuse.

As one can see from Fig. 3.69, resistance of the niobium oxide capacitors after breakdown of their dielectric depends on the fuse value limiting current in the circuit. There is broader distribution of the resistances with the mean value about 100 Ohm in the circuits with 5 A fuse and narrower distribution of the resistances with the mean value about 5 kOhm in the circuits with 1 A fuse. The resistance of the similar tantalum and niobium capacitors after BDV test is equal or below 1 Ohm regardless of the fuse value. Relatively high resistance after breakdown of the dielectric allows failed niobium oxide capacitors to stay in a "power on" circuit without overheating, especially, when applied voltage is reduced in comparison to the rated voltage.

The resistance of the capacitors after breakdown depends on the structure and composition of the breakdown channel that forms in the dielectric when accumulated energy is rapidly released through a small fault site in the dielectric causing instant temperature increase and material redistribution. Since in porous anodes most of the dielectric surface area is hidden in the pores, the investigation of the breakdown channels was performed on the flat models using sputtered metal films as anodes [18]. Figure 3.70 presents transmission electron microscopy image and electron diffraction pattern of the breakdown channel in the Nb_2O_5 dielectric formed on sputtered niobium film.

According to Fig. 3.70, the breakdown channel is a small polycrystalline area inside the amorphous matrix of the Nb_2O_5 dielectric containing Nb, NbO, and NbO_2 phases. In case of NbO anodes, the semiconductor NbO_2 phase with higher resistivity in comparison to the resistivity of Nb and NbO phases defines the resistance of the breakdown channel in the Nb_2O_5 dielectric. The relative amount of the NbO_2 phase and, thereby, resistance of the capacitor after breakdown depends on the

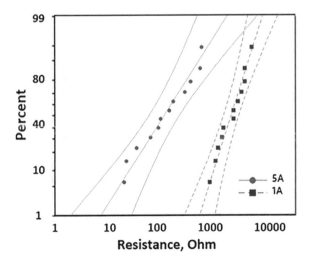

Fig. 3.69 Resistance of the D-case 150 μF–10 V niobium oxide capacitors after BDV test with either 1 or 5 A fuse

Fig. 3.70 Transmission
electron microscopy image
and electron diffraction
pattern of the breakdown
channel in the Nb_2O_5
dielectric

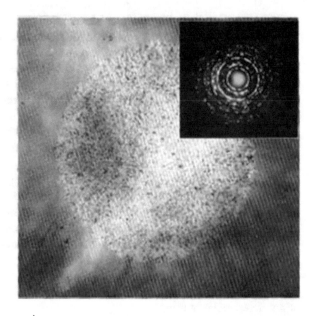

capacitor energy $E = CV^2/2$ and amperage of the fuse that limits current in the circuit during the BDV test.

A characteristic feature of the flat models with the NbO_2 phase in the breakdown channel in the Nb_2O_5 dielectric is the area of negative differential resistance on the current-voltage characteristic (Fig. 3.71) [18].

A similar current-voltage characteristic with the area of negative differential resistance was detected in the niobium oxide capacitor with porous NbO anode after the BDV test [21]. The similarities in current-voltage characteristics between the flat models and actual capacitors indicate that the breakdown channel with the NbO_2 phase also forms in failed niobium oxide capacitors and determines relatively high resistance of these capacitors after breakdown of the Nb_2O_5 dielectric.

The no short failure mode in niobium oxide capacitors is especially important in higher-voltage capacitors with larger accumulated energy; however, niobium oxide capacitors have lower working voltages in comparison to niobium and especially tantalum capacitors. At the same time, more stable behavior and higher working voltages are expected in niobium oxide capacitors in comparison to niobium capacitors since according to the niobium-oxygen equilibrium diagram, NbO-Nb_2O_5 interface is thermodynamically more stable than Nb-Nb_2O_5 interface [23]. Besides that, pure NbO phase doesn't promote crystallization of the amorphous matrix of the Nb_2O_5 dielectric [19].

A possible degradation mechanism that affects stability and limits working voltages in niobium oxide capacitors can be oxygen redistribution between the NbO powder with 50% atm of oxygen and practically free of oxygen tantalum or niobium lead wire embedded in the NbO powder at pressing. The oxygen flow increases exponentially with temperature and becomes very active at sintering temperature

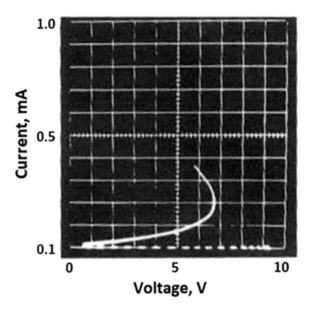

Fig. 3.71 Current-voltage characteristic of breakdown channel with NbO_2 phase

equal to or exceeding 1200 °C. Because of the oxygen redistribution between the powder and the lead wire, tantalum or niobium lead wire becomes enriched with oxygen, while depleted with oxygen, NbO powder in vicinity of the lead wire undergoes phase transforms into niobium saturated with oxygen with inclusions of the NbO phase [54]. The amorphous matrix of the Nb_2O_5 dielectric in this area of the NbO anodes is highly susceptible to the field crystallization, especially, when formation voltage increases to form thicker dielectric for the higher working voltage capacitors. Usage of the NbO lead or a high density NbO nob on the NbO anodes eliminates the oxygen redistribution between the powder and the lead and allows for higher working voltages in the niobium oxide capacitors [145].

In conclusion, niobium-based capacitors provide lower cost per *CV* solution in comparison to tantalum capacitors due to the larger availability and lower cost of niobium in comparison to tantalum. Besides that, niobium oxide capacitors provide safety feature such as no short failure mode. The relatively high resistance of the failed niobium oxide capacitors is due to the presence of the semiconductor NbO_2 phase in the breakdown channel in the Nb_2O_5 dielectric. In comparison to tantalum capacitors, niobium and niobium oxide capacitors have larger size for given *CV* and lower working voltages, especially, in case of the niobium oxide capacitors, where working voltages are limited to 10 V. The phase transformations in NbO anodes in vicinity of the embedded tantalum or niobium lead wire during the anode sintering activate field crystallization of the amorphous matrix of the Nb_2O_5 dielectric and limit formation voltages and, thereby, working voltages of the niobium oxide capacitors.

Chapter 4
Conclusion

Tantalum capacitors have been on the market for three-quarters of the century. The major advantage of tantalum capacitors in comparison to other major types of capacitors, such as multilayer ceramic capacitors, film capacitors, and aluminum electrolytic capacitors, is their record high volumetric charge and energy efficiency. The high efficiency is provided by the large surface area per unit of volume of tantalum anodes sintered with fine tantalum powder and small submicron thickness of the anodic oxide film of tantalum employed as a dielectric in tantalum capacitors. The other advantage of tantalum capacitors in comparison to other types of the capacitors is capacitance stability with voltage and temperature. There is practically no change in capacitance measured with and without DC bias and small variation in capacitance within normal range of operating temperatures. Besides that, there is no aging of tantalum capacitors; their capacitance remains stable during any practical duration of the testing and field application.

Tantalum capacitors evolved from Wet Tantalum capacitors with liquid electrolyte cathode to Solid Electrolytic Tantalum capacitors with MnO_2 cathode and then to Polymer Tantalum capacitors with conductive polymer cathode. Higher conductivity of the MnO_2 in comparison to the liquid electrolyte and conductive polymer in comparison to the MnO_2 provided substantial reduction in equivalent series resistance and, thereby, better capacitance stability with frequency and improved ripple current capability of tantalum capacitors. At the same time, the evolution of tantalum capacitors from Wet to Solid Electrolytic and then to polymer was accompanied by significant reduction in working voltages and an increase in DC leakage at every step of the evolution. When Polymer Tantalum capacitors were introduced to the market, their applications were limited to low-voltage commercial electronics.

The most important improvements in performance and reliability of all types of tantalum capacitors were achieved due to the suppressing of the degradation processes in the Ta_2O_5 dielectric and its interface with tantalum anode. These degradation processes, oxygen migration from the dielectric into the anode and crystallization of the amorphous matrix of the dielectric, have roots in the thermodynamic

Y. Freeman, *Tantalum and Niobium-Based Capacitors*,
https://doi.org/10.1007/978-3-030-89514-3_4

instability of the Ta-Ta_2O_5 bilayer and are the major reason for the parametric and catastrophic failures of tantalum capacitors. Flawless dielectric technology (F-Tech) was developed, which allowed the manufacturing of tantalum capacitors with the Ta_2O_5 dielectric practically free from structural defects in absolute majority of the population of capacitors. Nevertheless, as with any technology, there was a probability of a small percentage of the potentially unreliable parts with hidden defects in the dielectric that were not detectable by the existing testing techniques. To address this issue, simulated breakdown screening (SBDS) was developed that identifies and removes the parts with hidden defects in the dielectric without any damage to the entire population of capacitors.

The F-Tech and SBDS provide fundamental improvement in reliability of tantalum capacitors and allow their application with low or no de-rating. Solid Electrolytic Tantalum capacitors manufactured with F-Tech and SBDS have significantly lower failure rate in comparison to the failure rate in the capacitors manufactured with conventional technology. Polymer Tantalum capacitors manufactured with F-Tech have the lowest failure rate, which is decreasing with time of the long-term testing similar to that in Solid Electrolytic Tantalum capacitors.

There is no ignition and burning tantalum in failed short SMD-type Solid Electrolytic and Polymer Tantalum capacitors. The black marks typically observed on the surface of the failed surface-mount tantalum capacitors are the areas of the epoxy compound carbonized under the heat propagated from the fault sites in the dielectric at the breakdown event. The epoxy compounds used in manufacturing of SMD-type tantalum capacitors and other types of the electronic components is flame retardant, meets high safety standard and usually don't ignite in the capacitors failed at normal conditions of the accelerated testing.

The lower-voltage Polymer Tantalum capacitors allow usage of the highest CV/g tantalum powders, and, thereby, further increase in their charge efficiency. At the same time, in lower-voltage Solid Electrolytic Tantalum capacitors, the thermal oxide grows under the thin anodic oxide of tantalum during the pyrolytic deposition of the MnO_2 cathode, causing dielectric thickness increase and capacitance loss. This process is intensified by the high oxygen content in tantalum anodes sintered with high CV/g tantalum powder.

In the middle range of working voltages, a further increase in energy efficiency can be achieved through flawless dielectric technology with tantalum anodes sintered in deoxidizing atmosphere (F-Tech with deox-sintering). These anodes have a low oxygen content and a unique morphology combining thick necks connecting the powder particles and large open pores between these particles. F-Tech with deox-sintering allows manufacturing of Polymer Tantalum capacitors with highest volumetric efficiency and lowest failure rate at the long-term testing. Progress in packaging also contributes to further increase in charge and energy efficiency of tantalum capacitors. The packaging principles are not included in this book as they are common for a broad range of the electronic components.

In the high voltage range of working voltages, significant progress in performance and reliability of Polymer Tantalum capacitors was achieved by the combination of F-Tech in anode and dielectric manufacturing with pre-polymerized slurry

PEDOT technology in cathode manufacturing. These technologies provide an efficient potential barrier at the dielectric interface with the polymer cathode exhibiting p-type semiconductor properties. At normal polarity (+ on tantalum anode), the barrier at the Ta_2O_5-polymer interface increases with DC voltage, limiting current through the capacitor. The F-Tech and slurry polymer technologies provided a foundation for development of Polymer Tantalum capacitors with record high working voltage and record low DC leakage never achieved before in any Solid Tantalum capacitors and comparable to these in Wet Tantalum capacitors. Low-temperature manufacturing, which prevents thermal crystallization of the thicker Ta_2O_5 dielectric, strong self-healing properties of the slurry PEDOT cathode, and simulated breakdown screening also contributed to the outstanding performance and reliability of the high-voltage Polymer Tantalum capacitors. These capacitors are now broadly used in the most demanding high-reliability applications, such as military and aerospace, and their development continues aiming at larger CV, lower ESR, and better mechanical robustness.

The impact of the environmental conditions on stability of the DC and AC characteristics of the Solid Tantalum capacitors was investigated in broad range of working voltages. While the environmental impact on stability of the DC characteristics, such as DCL and BDV, is more pronounced in higher-voltage tantalum capacitors with thicker dielectrics more prone to crystallization, the environmental impact on stability of the AC characteristics, such as capacitance change with temperature $C(T)$ and frequency $C(f)$, is more pronounced in lower-voltage tantalum capacitors with thinner dielectrics. The latter can be related to the changes in the morphology of the dielectric surface with increasing thickness of the anodic Ta_2O_5 film, where the surface roughness decreases for thicker films but increases for thinner films. These changes in morphology of the dielectric surface affect the coverage of the dielectric with MnO_2 and polymer cathode and, thereby, the capacitance and its stability with temperature and frequency.

The coverage of the dielectric surface with pre-polymerized PEDOT particles in Polymer Tantalum capacitors with slurry PEDOT cathode plays a critical role in the capacitance stability in these capacitors. Under humid conditions, the conductivity of the dielectric surface increases in the free area where there is no direct contact between the dielectric and polymer. This increase in surface conductivity of the dielectric, provided by the thin layer of absorbed water molecules, results in a higher total surface area and, thereby, a higher capacitance in humid parts in comparison to the capacitance in dry parts. At lower temperatures and higher frequencies, the conductivity of the humidified dielectric surface decreases due to the lower mobility of the ionic charge carriers, which increases capacitance loss with temperature and frequency. Improving the integrity of the dielectric-polymer interface (lowering the free area) reduces the effect of the environmental conditions and improves capacitance stability in Polymer Tantalum capacitors.

The anomalous currents at normal polarity (+ on the tantalum anode) including anomalous transient current when a pulse of rated voltage is applied to the capacitor, anomalous charge current when rapid voltage ramp is applied to the capacitor, and also anomalous DC current when rated voltage is applied to the capacitors at low temperature, were detected in dry Polymer Tantalum capacitors with hybrid in situ—slurry and pure slurry PEDOT cathode. Anomalous currents decrease with the humidification and pulse repetitions, but they can be restored by drying the capacitors or by application of the voltage pulse at reverse polarity (– on tantalum anode).

The anomalous currents observed in Polymer Tantalum capacitors can be explained by the presence of dipoles, charged polymer chains, in the conducting polymer cathode at its interface with the oxide dielectric. These dipoles will reorient when an electrical field is applied, and this reorientation determines the potential barrier at the dielectric-polymer interface, which plays a crucial role in limiting the current through the Polymer Tantalum capacitor. The dipole orientation time is a function of temperature, and therefore the process slows down at lower temperatures when mobility of polymer chains becomes low. In humidified capacitors the reorientation is faster than the process in dry capacitors, since the water acts as a plasticizer for the polymer chains. This reorientation, and the resulting anomalous currents, is more pronounced in pre-polymerized slurry PEDOT cathodes with long molecules of PSS dopant in the polymer structure; therefore, it is more pronounced in capacitors with the hybrid or pure slurry polymer cathodes as compared to those with pure in situ polymer cathodes. In the higher-voltage Polymer Tantalum capacitors with slurry PEDOT cathodes quality of the tantalum oxide dielectric also plays important role in the anomalous currents. Particularly flawless dielectric technology (F-Tech) provides significant reduction in anomalous charge current when rapid voltage ramp is applied to the capacitor after thorough drying in comparison to that in Polymer Tantalum capacitors manufactured with conventional dielectric technology.

There are no anomalous currents in Solid Electrolytic Tantalum capacitors at normal operating condition. The additional advantage of Solid Electrolytic Tantalum capacitors over Wet and Polymer Tantalum capacitors is their ability to work in harsh environment, particularly high operating temperatures. These high operating temperatures are not achievable in Wet Tantalum capacitors due to the pressure increase in the hermetic can with liquid electrolyte cathode and in Polymer Tantalum capacitors due to the glass transition and decomposition of the polymer cathode. Similar to the high-reliability technology, high-temperature technology is focused on suppressing the degradation processes in the Ta_2O_5 dielectric and its interfaces with the anode and cathode, which activate rapidly at temperatures exceeding 200 °C. The F-Tech is an important part of the high-temperature technology. Manufacturing of the high-temperature Solid Electrolytic Tantalum capacitors also includes thermally stable materials in external metallization, assembly, and encapsulation. With these technologies and materials, operating temperatures in Solid Electrolytic Tantalum capacitors can be increased to 250 °C or even higher temperatures.

Niobium and niobium oxide capacitors with niobium (Nb) or niobium monoxide (NbO) anodes, Nb_2O_5 dielectric, and MnO_2 cathodes provide lower cost per CV solution in comparison to tantalum capacitors. Niobium oxide capacitors also provide safety feature such as no short failure mode. At the same time, niobium-based capacitors have lower volumetric efficiency in comparison to tantalum capacitors, which is partially due to a much finer tantalum powders used in manufacturing of the tantalum capacitors in comparison to the coarser niobium and niobium monoxide powders used in manufacturing of the niobium-based capacitors. Furthermore, more active degradation processes in the Nb_2O_5 dielectric in comparison to the Ta_2O_5 dielectric limit formation voltages and, thereby, working voltages in niobium-based capacitors. Niobium oxide capacitors have the lowest working voltages even though their basic bilayer NbO-Nb_2O_5 is thermodynamically more stable than basic bilayer Nb-Nb_2O_5 in niobium capacitors. Working voltages in niobium oxide capacitors can be limited by phase transformations in the NbO anodes in the vicinity of the embedded in these anodes tantalum or niobium lead wire. These phase transformations take place during sintering of the NbO anodes and are originated by the active redistribution of oxygen between the NbO particles with 50% atm of oxygen and tantalum or niobium lead wire that is practically free of oxygen. Using NbO lead in the NbO anodes will eliminate the oxygen redistribution during sintering of the NbO anodes and allow higher working voltages in niobium oxide capacitors.

References

1. C. Chaneliere, J.L. Autran, R. Devine, B. Balland, Mater. Sci. Eng. **R22**, 269 (1998)
2. D.F. Untereker, C.L. Schmidt, G. Jain, P.A. Tamirisa, J.H. Schott, M. Viste, in *Clinical Cardiac Pacing, Defibrillation, and Resynchronization Therapy*, ed. by K. A. Ellenbogen, B. L. Wilkoff, G. N. Kay, C. P. Lau, A. Auricchio 5th, (Elsevier, Philadelphia, 2016), pp. 251–269
3. T.B. Tripp, Tantalum and Tantalum compounds, in *Kirk-Othmer Encyclopedia of Chemical Technology*, 4th edn., (Wiley, Hoboken, 1997)
4. J. Bardeen, W.H. Brattain, The transistor, a semi-conductor triode. Phys. Rev. **74**, 230 (1948)
5. H.E. Haring, N. Summit, R.L. Taylor, U.S. Patent 3,166,693
6. D.M. Smyth, G.A. Shirn, T.B. Tripp, J. Electrochem. Soc. **110**, 1264 (1963)
7. D.M. Smyth, T.B. Tripp, J. Electrochem. Soc. **110**, 1271 (1963)
8. D.M. Smyth, G.A. Shirn, T.B. Tripp, J. Electrochem. Soc. **111**, 1331 (1964)
9. D.M. Smyth, T.B. Tripp, G.A. Shirn, J. Electrochem. Soc. **113**, 101 (1966)
10. D.M. Smyth, J. Electrochem. Soc. **113**(3), 1324 (1966)
11. D.M. Smyth, J. Electrochem. Soc. **114**, 723 (1967)
12. D.M. Smyth, G.A. Shirn, J. Electrochem. Soc. **115**, 186 (1968)
13. A. Kobayashi, T. Nishiyama, K. Watanabe, T. Nakata, K. Morimoto, NEC Res. Dev. **32**, 66 (1984)
14. S. Kirchmeyer, K. Reuter, J. Mater. Chem. **15**, 2077 (2005)
15. N. Koch, A. Vollmer, A. Elschner, Appl. Phys. Lett. **90**, 043512 (2007)
16. K. Ueno, L. Dominey, R. Alwitt, in *Proceedings of the 211th Meeting of The Electrochemical Society–B1-Electrochemistry of Novel Electrode Materials for Energy Conversion and Storage*, Chicago, 2007
17. L. Young, *Anodic Oxide Films* (Academic Press, New York, 1961)
18. B. Boiko, P. Pancheha, V. Kopach, Y. Pozdeev-Freeman, Thin Solid Films **130**, 341 (1985)
19. B. Boiko, V. Kopach, S. Melentyev, P. Pancheha, Y. Pozdeev-Freeman, V. Starikov, Thin Solid Films **229**, 207 (1993)
20. J. Fife, U.S. Patent 6,391,275 B2
21. S. Zednichek, Z. Sita, T. Zednichek, M. Komarek, in *Proceedings of 24th Capacitor and Resistor Technology Symposium (CARTS)*, San Antonio, 2004, p. 223
22. Y. Pozdeev-Freeman, D. Johnston, D. Wadler, in *Proceedings of 23rd Capacitor and Resistor Technology Symposium (CARTS)*, Scottsdale, 2003, p. 47
23. E. Fromm, E. Hebhardt, *Gase und Kohlenstaff in Metallen* (Springer, Berlin, 1976)
24. B. Predel, in *Phase Equilibria, Crystallographic and Thermodynamic Data of Binary Alloys*, ed. by O. Madelung, (Springer International, Berlin/London, 1998)

© The Author(s), under exclusive license to Springer Nature Switzerland AG 2022
Y. Freeman, *Tantalum and Niobium-Based Capacitors*,
https://doi.org/10.1007/978-3-030-89514-3

25. Y. Pozdeev-Freeman, A. Gladkikh, M. Karpovski, A. Palevsky, J. Electron. Mater. **27**, 1034 (1998)
26. L. Yang, M. Viste, J. Hossick-Schott, B. Sheldon, Electrochim. Acta **81**, 90 (2012)
27. D.A. Vermilyea, Acta Metall. **5**, 113 (1957)
28. D.A. Vermilyea, J. Electrochem. Soc. **104**, 485 (1957)
29. J.J. Randell, W.J. Bernard, R.R. Wilkinson, Electrochim. Acta **10**, 183 (1965)
30. J.J. Randell, Electrochim. Acta **20**, 63 (1975)
31. T.B. Tripp, R.M. Creasi, B. Cox, in *Proceedings of 20th Capacitor and Resistor Technology Symposium (CARTS)*, Huntington Beach, 2000, p. 256
32. G.P. Klein, J. Electrochem. Soc. **119**, 1551 (1972)
33. R.E. Pawel, J. Electrochem. Soc. **114**, 1222 (1967)
34. W. Anders, Thin Solid Films **27**, 135 (1975)
35. M.H. Rottersman, M.J. Bill, D. Gertstenberg, IEEE Compon. **1**, 137 (1976)
36. P. Wyatt, IEEE Compon. **1**, 148 (1978)
37. P.K. Reddy, S.R. Jawalehan, Thin Solid Films **64**, 71 (1979)
38. T.B. Tripp, M. Shaw, B. Cox, *Proceedings of 19th Capacitor and Resistor Technology Symposium (CARTS)*, vol 19 (New Orleans, 1999), p. 317
39. R. Hahn, B. Melody, J. Kinard, D. Wheeler, U.S. Patent 6,214,271 B1
40. R.E. Pawel, J.P. Pensler, C.A. Evans, J. Electrochem. Soc. **119**, 24 (1972)
41. N.F. Jackson, J. Appl. Electrochem. **3**, 91 (1973)
42. G.E. Cavigliasso, M.J. Esplandiu, V.A. Macagano, J. Appl. Electrochem. **28**, 1213 (1998)
43. Y.M. Li, L. Young, J. Electrochem. Soc. **147**, 1344 (2000)
44. B. Melody, T. Kinard, P. Lessner, in *Proceedings of 19th Capacitor and Resistor Technology Symposium (CARTS)*, New Orleans, 1999, p. 84
45. G.P. Klein, J. Electrochem. Soc. **113**, 348 (1966)
46. Y. Pozdeev-Freeman, A. Gladkikh, J. Electron. Mater. **30**, 931 (2001)
47. M. Tierman, R.J. Millard, in *Proceedings of 33rd Electronic Components Conference*, Orlando, 1983
48. Y. Pozdeev-Freeman, Y. Rozenberg, A. Gladkikh, M. Karpovski, A. Palevski, J. Mater. Sci. Mater. Electron. **9**, 309 (1998)
49. Y. Freeman, P. Lessner, A.J. Kramer, J. Li, E.C. Dickey, J. Koenitzer, L. Mann, Q. Chen, T. Kinard, J. Qazi, J. Electrochem. Soc. **157**(7), G161 (2010)
50. R.E. Pawel, J.J. Campbell, J. Electrochem. Soc. **111**, 1230 (1964)
51. G.V. Samsonov, *Handbook of Physicochemical Properties of Oxides* (Metallurgia, Moscow, 1978)
52. G.P. Klein, Proc. IEEE **1**, 70 (1965)
53. Y. Pozdeev-Freeman, Qual. Reliab. Eng. Int. **14**, 79 (1998)
54. Y. Pozdeev-Freeman, A. Gladkikh, Y. Rosenberg, Mater. Res. Soc. Symp. Proc. **788**, L3.32 (2004)
55. Y. Pozdeev-Freeman, P. Maden, in *Proceedings of 22nd Capacitor and Resistor Technology Symposium (CARTS)*, New Orleans, 2002, p. 148
56. I.M. Robertson, P. Sofronis, A. Nagao, M.L. Martin, S. Wang, D.M. Gross, K.E. Nygren, *Edward DeMille Campbell Memorial Lecture* (ASM International, Novelty, OH, 2014)
57. Y. Freeman, P. Lessner, R. Hahn, J. Prymak, *Passive Component Industry* (ECA, 2007), p. 22
58. H. Haas, M. Hagymasi, H. Brumm, C. Schnitter, U.S. Patent 9,378,894 B2
59. T.B. Tripp, J. Eckert, *Kirk-Othmer Encyclopedia of Chemical Technology*, vol 23, 4th edn. (Wiley, Hoboken, 1977), p. 658
60. D.L. Perry, *Handbook of Inorganic Compounds*, 2nd edn. (CRC, Boca Raton, 2011)
61. L. Shekhter, T. Tripp, L. Lanin, A. Conlon, H. Goldberg, U.S. Patent 6,849,104 B2
62. A. Michaelis, C. Schnitter, K. Reichert, R. Wolf, U. Merker, in *Proceedings of Capacitor and Resistor Technology Symposium (CARTS)*, New Orleans, 2002, p. 209
63. J. Satterfield, L. Thornton, J. Poltorak, R. Hahn, Y. Qiu, U.S. Patent 7,116,548
64. R. Hahn, J. Kinard, J. Poltorak, E. Zediak, U.S. Patent 7,154,742

65. G.C. Kuczynski, Physics and chemistry of sintering. Adv. Colloid Interf. Sci. **33**(3), 275 (1972)
66. Y. Freeman, P. Lessner, In *Passive Components Industry*, July/August 2008, p. 22
67. B.J. Melody, J.T. Kinard, D.A. Wheeler, U.S. Patent 5,716,511
68. J.T. Kinard, B.J. Melody, D.A. Wheeler, U.S. Patent 6,480,371 B1
69. L.L. Odynets, Sov. Electrochem. **23**, 1703 (1987)
70. L.L. Odynets, V.M. Orlov, *Anodic Oxide Films* (Nauka, Leningrad, 1990) (in Russian)
71. A.F. Torrisi, J. Electrochem. Soc. **102**, 176 (1955)
72. B.J. Melody, J.T. Kinard, P.M. Lessner, U.S. Patent 6,261,434 B1
73. D. Evans, U.S. Patent 5,469,325 A
74. D. Evans, U.S. Patent 5,737,181 A
75. www.Vishay Intertechnology. Wet Electrolyte Tantalum Capacitors
76. A. Scotheim, J. R. Reynolds (eds.), *Handbook of Conducting Polymers*, 3rd edn. (CRC, Boca Raton, 2006)
77. Y. Kedoh, K. Akami, H. Kusayanagi, Y. Matsuya, Synth. Met. **123**, 541 (2001)
78. B.W. Jensen, K. West, Macromolecules **37**, 4538 (2004)
79. Y. Qiu, R.S. Hahn, K.R. Brenneman, U.S. Patent 7,563,290 B2
80. U. Merker, W. Lovenich, K. Wussow, U.S. Patent 8,313,538 B2
81. Y. Freeman, W.R. Harrell, I. Luzinov, B. Holman, P. Lessner, J. Electrochem. Soc. **156**(6), G65 (2009)
82. J.W. Saterfield, L.P. Thornton, J.P. Poltorak, R. Hahn, Y. Qiu, U.S. Patent 7,116,548 B2
83. A. Chacko, J. Young, R. Hahn, in *Proceedings of 29th Annual Passive Components Symposium* , Jacksonville, 2009, p. 265
84. B. Melody, T. Kinard, K. Moore, D. Wheeler, U.S. Patent 6,319,459 B1
85. Y. Freeman, U.S. Patent 7,761,603 B2
86. J. Paulsen, E. Reed, Y. Freeman, U.S. Patent 8,441,265 B2
87. Y. Freeman, *Passive Component Industry* (Electronic Industries Alliance, Paumanok Publications, January/February 2005), p. 6
88. B. Melody, T. Kinard, D. Wheeler, *Proceedings of the 21st Capacitor and Resistor Technology Symposium (CARTS)* (St. Petersburg, 2001), p. 57
89. L. Simkins, M.J. Albarelli, K.B. Doyle, B.L. Cox, *Proceedings of the Capacitor and Resistor Technology Symposium (CARTS)* (San Antonio, 2004), p. 47
90. Y. Freeman, P.M. Lessner, J. Poltorak, S.C. Hussey, U.S. Patent 8,308,825 B2
91. W. Albrecht, A. Hoppe, U. Papp, R. Wolf, U.S. Patent 4,537,641
92. Y. Pozdeev-Freeman, *Proceedings International Symposium on Tantalum and Niobium* (San Francisco, 2000), p. 291
93. Y. Pozdeev-Freeman, U.S. Patent 5,825,611
94. Y. Pozdeev-Freeman, U.S. Patent 6,447,570 B1
95. Y. Freeman, P. Lessner, U.S. Patent 7,731,803 B1
96. S.C. Hussey, Y. Freeman, P.M. Lessner, U.S. Patent 8,349,030 B1
97. C. Guerrero, J. Poltorak, Y. Freeman, S. Hussey, C. Stolarski, U.S. Patent 10,290,429 B2
98. http://www.landandmaritime.dla.mil/Downloads/MilSpec/Docs/MIL-PRF-55365/prf55365.pdf
99. J. Paulsen, E. Reed, *1st Capacitor and Resistor Technology Symposium (CARTS)* (St. Petersburg, FL, 2001), p. 265
100. J. Paulsen, E. Reed, J. Kelly, *24th Capacitor and Resistor Technology Symposium (CARTS)* (San Antonio, TX, 2004), p. 114
101. J. Prymak, *21st Capacitor and Resistor Technology Symposium (CARTS)* (St. Petersburg, FL, 2001), p. 25
102. J. Prymak, *23rd Capacitor and Resistor Technology Symposium (CARTS)* (Scottsdale, AZ, 2003), p. 278
103. Y. Freeman, P. Lessner, I. Luzinov, J. Solid State Sci. Technol. **10**, 045007 (2021)
104. W. Winkel, E. Rich, *34th Symposium for Passive Electronic Components* (Santa Clara, CA, 2014), p. 175

105. Y. Freeman, P. Lessner, Appl. Sci. **11**(12), 5514. Special Issue Multifanctional Polymers and Composites (2021)
106. G. Camino, G. Tartaglione, A. Frache, C. Manferti, G. Costa, Polym. Degrad. Stab. **90**(2), 354–362 (2005)
107. S.V. Levchik, G. Camino, M.P. Luda, L. Costa, G. Muller, B. Costes, Polym. Degrad. Stab. **60**(1), 169–183 (1998)
108. S.V. Levchik, G. Camino, M.P. Luda, L. Costa, G. Muller, B. Costes, Y. Henry, Polym. Adv. Technol. **7**(11), 823–830 (1996)
109. S.V. Levchik, G. Camino, L. Costa, M.P. Luda, Polym. Degrad. Stab. **54**(2–3), 317–322 (1996)
110. Y. Freeman, G. Alapatt, W. Harrell, P. Lessner, J. Electrochem. Soc. **159**(10), A1646 (2012)
111. Y. Freeman, Y. Qiu, S. Hussey, P. Lessner, U.S. Patent 8,310,815 B2
112. A. Gurav, X. Xu, Y. Freeman, E. Reed, KEMET electronics: breakthroughs in Ceramic and Tantalum Capacitor Technology, in *Materials Research for Manufacturing, Springer Series in Materials Science*, ed. by L. Madsen, E. Svedberg, vol. 224, (Springer International, Cham, 2016), p. 93
113. Y. Freeman, S. Hussey, J. Chen, T. Kinard, E. Jones, H. Bishop, H. Perkins, K. Tempel, E. Reed, J. Paulsen, *Proceedings Quality and Reliability Technical Symposium (QRTS)* (Electronic Component Industry Association, Mesa, 2015), p. 7
114. Y. Freeman, J. Chen, R. Fuller, S. Hussey, E. Jones, T. Kinard, P. Lessner, M. Maich, T. McKinney, *Proceedings CARTS-Europe* (ECA, Munich, 2010), p. 143
115. A. Lenz, H. Kariis, A. Pohl, P. Persson, L. Ojamae, Chem. Phys. **384**, 44 (2011)
116. Y. Saiki, T. Nakata, NEC Res. Dev. **32**(3), 332 (1991)
117. D.A. Vermilyea, J. Appl. Phys. **36**(11), 3663 (1965)
118. N. Axelrod, N. Schwartz, J. Electrochem. Soc. **116**(4), 460 (1969)
119. R.M. Fleming et al., J. Appl. Phys. **88**(2), 850 (2000)
120. A. Teverovsky, in *Proceedings Capacitors and Resistors Technology Symposium (CARTS)*, New Orleans, 2002, p. 105
121. A. Teverovsky, in *Proceedings Capacitors and Resistors Technology Symposium (CARTS)*, Jacksonville, 2011, p. 161
122. Y. Sasaki, J. Phys. Chem. Solids **13**, 177 (1960)
123. J.D. Prymak, in *Proceedings Capacitors and Resistors Technology Symposium*, New Orleans, 2002, p. 101
124. Y. Freeman, G.F. Alapatt, W.R. Harrell, I. Luzinov, P. Lessner, ECS J. Solid State Sci. Technol. **4**(7), N70 (2015)
125. Q. Chen, Y. Freeman, S. Hussey, U.S. Patent 8,379134,371 B2, 2013
126. Y. Freeman, I. Luzinov, R. Burtovyy, P. Lessner, W.R. Harrell, S. Chinnam, J. Qazi, ECS J. Solid State Sci. Technol. **6**(7), N104 (2017)
127. E.N. Tarekegn, W.R. Harrell, I. Luzinov, P. Lessner, Y. Freeman, ECS J. Solid State Sci. Technol. **9**, 083005 (2020)
128. I. Abuetwirat, *Dielectric Properties of Thin Tantalum and Niobium Oxide Layers*, Doctor Thesis, Brno University of Technology, Brno, 2014
129. M. Scholdt, H. Do, J. Lang, A. Gall, A. Colsmann, U. Lemmer, J.D. Koenig, M. Winkler, H. Boettner, J. Electron. Mat. **39**(9), 1589 (2010)
130. I. Horacek, T. Zednicek, M. Komarec, J. Tomasco, S. Zednicek, W.A. Millman, J. Sikula, J. Hlavka, in *Proceedings of the 22nd CARTS, New Orleans, LA, 26–28 March* 2002
131. W.C.D. Dampier, *Proceedings of the Royal Society of London. Series A, Containing Papers of a Mathematical and Physical Character*, vol 76(513) (1905), pp. 577–583
132. A. Teverovsky, IEEE Trans. Comp. Packag. Manuf. Technol. **9**, 11 (2019)
133. U. Merker, K. Asteman, European Patent No. 2,622,616 B1
134. Y. Freeman, G.F. Alapatt, W.R. Harrell, I. Luzinov, P. Lessner, J. Qazi, ECS J. Solid State Sci. Technol. **2**(11), N197 (2013)
135. L.H. Sperling, *Introduction to Physical Polymer Science* (Wiley, Hoboken, 2006)
136. Y. Freeman, S. Hussey, J. Cisson, P. Lessner, U.S. Patent 10,062,519 B2

137. M.R. Werner, W.R. Fahrner, IEEE Trans. Ind. Electron. **48**(2), 249 (2001)
138. R.W. Johnson, J.L. Evans, P. Jacobsen, J.R. Thompson, M. Christopher, IEEE Trans. Electron. Packag. Manuf. **27**(3), 164 (2004)
139. R. Hahn, K. Tempel, *International Conference and Exhibition on High Temperature Electronics Network (HiTEN)* (Cambridge, 2015)
140. A. Cashion, G. Cieslewski, *International Conference and Exhibition on High Temperature Electronics Network (HiTEN)* (Cambridge, 2015)
141. A. Chacko, U.S. Patent 8,310,816 B2
142. K. Reichelt, X. Jiang, Thin Solid Films **191**(1), 91 (1990)
143. J. McConnell, J. Bultitude, R. Phillips, R. Hill, G. Renner, P. Lessner, A. Chaco, J. Bell, K. Brown, U.S. Patent 8,902,565 B2
144. A. Chaco, J. McConnell, P. Lessner, R. Hahn, J. Bultitude, U.S. Patent 8,896,986 B2
145. Y. Freeman, P. Lessner, J. Poltorak, R. Hahn, U.S. Patent 8,825,465 B2

Index

Printed in the United States
by Baker & Taylor Publisher Services